Stude
Solutions Manual

Interactive
Statistics
Second Edition

MARTHA ALIAGA

BRENDA GUNDERSON

Prentice
Hall

Upper Saddle River, NJ 07458

Editor in Chief: Sally Yagan
Supplement Editor: Joanne Wendelken
Assistant Managing Editor: John Matthews
Production Editor: Donna Crilly
Supplement Cover Manager: Paul Gourhan
Supplement Cover Designer: Joanne Alexandris
Manufacturing Buyer: Ilene Kahn

© 2003 by Pearson Education, Inc.
Pearson Education, Inc.
Upper Saddle River, NJ 07458

All rights reserved. No part of this book may be reproduced in any form or by any means, without permission in writing from the publisher.

The author and publisher of this book have used their best efforts in preparing this book. These efforts include the development, research, and testing of the theories and programs to determine their effectiveness. The author and publisher make no warranty of any kind, expressed or implied, with regard to these programs or the documentation contained in this book. The author and publisher shall not be liable in any event for incidental or consequential damages in connection with, or arising out of, the furnishing, performance, or use of these programs.

Printed in the United States of America

10 9 8 7 6 5 4 3 2

ISBN 0-13-065846-4

Pearson Education Ltd., *London*
Pearson Education Australia Pty. Ltd., *Sydney*
Pearson Education Singapore, Pte. Ltd.
Pearson Education North Asia Ltd., *Hong Kong*
Pearson Education Canada, Inc., *Toronto*
Pearson Educacíon de Mexico, S.A. de C.V.
Pearson Education—Japan, *Tokyo*
Pearson Education Malaysia, Pte. Ltd.
Pearson Education, *Upper Saddle River, New Jersey*

Contents

Interactive Statistics 2nd Edition: Chapter 1 Odd Solutions

1.1
In hypothesis testing, the purpose is to determine whether there is sufficient evidence with which to reject the null hypothesis (H_0), which generally reflects the prevailing viewpoint. The alternative hypothesis (H_1) is often what someone sets out to prove.

1.3
H_0: The 5-year survival rate for those using the vaccine is equal to 10%.
H_1: The 5-year survival rate for those using the vaccine is greater than 10%.

1.5
(a) Since a Type I error is rejecting H_0 when H_0 is true, we would be concluding the gun is not loaded when it is loaded. Since a Type II error is accepting H_0 when H_1 is true, we would be thinking the gun is loaded when it is not. A Type I error may be more serious as one might accidentally shoot a loaded gun.
(b) Since a Type I error is rejecting H_0 when H_0 is true, we would be concluding the dog does not bite when it does bite. Since a Type II error is accepting H_0 when H_1 is true, we would be thinking the dog bites when it does not. A Type I error may be more serious as we might approach a dog that could bite.
(c) Since a Type I error is rejecting H_0 when H_0 is true, we would be concluding the mall is closed when it is open. Since a Type II error is accepting H_0 when H_1 is true, we would be thinking the mall is open when it is closed. A Type II error may be more serious as we might waste the time and gas to drive to the mall expecting it open when it is closed.
(d) Since a Type I error is rejecting H_0 when H_0 is true, we would be concluding the watch is waterproof when it is not. Since a Type II error is accepting H_0 when H_1 is true, we would be thinking the watch is not waterproof when it is waterproof. A Type I error may be more serious as we might ruin our watch if it gets wet.

1.7
H_0: The average tomato yield for Brand A fertilizer is the same as the tomato average yield for the more expensive Brand B fertilizer.
H_1: The average tomato yield for the more expensive Brand B fertilizer is greater than the average tomato yield for the Brand A fertilizer.
Type I Error: Spend more money on the Brand B fertilizer when it really is not better than the Brand A fertilizer regarding the average tomato yield. Type II Error: Continue to use the Brand A fertilizer when the Brand B fertilizer results in a higher tomato yield on average.

1.9
A Type I error is rejecting the null hypothesis when it is true. So the owner would conclude the patrons are older and the owner would spend the time and money to remodel, when the crowd is actually not older. The owner would have spent money unnecessarily and the remodeling may not appeal to some of the patrons, but in general, it is not a serious error.

1.11
(a) The null hypothesis was accepted.
(b) No, a complaint was not registered.
(c) Yes, a Type II error may have been made. The cans are thought to be containing the stated sodium content when actually they contain higher amounts of sodium on average.

1.13
If α is decreased then β will increase, so the possible value is 0.30.

1.15
(a) The significance level α is 2/30=0.067 and the level of β is 20/30=0.667.
(b) Decision Rule #2: Reject H_0 if the selected voucher is $\leq$ \$2 or is $\geq$ \$9. The significance level α is 6/30=0.20 and the level of β is 12/30=0.40. Enlarging the rejection region resulted in increasing the level of α from 0.067 to 0.20 while decreasing the level of β from 0.667 to 0.40.

1.17

No, we need a decision rule that states when we reject or accept H_0.

1.19

(a) False: $\alpha + \beta$ does not need to equal 1. The value of α is calculated under H_0 while the value of β is calculated under H_0.

(b) False: Type II error is the chance of accepting H_0 when H_0 is true.

(c) True.

(d) False: H_0 is rejected if the sample shows evidence against it.

(e) False: The sample size does not influence the alternative hypothesis. The alternative hypothesis H_0 can be one-sided no matter what is the sample.

1.21

(a) H_0: The shown box is Box A. H_1: The shown box is Box B.

(b) The direction of extreme is one-sided to the left.

(c) Reject H_0 if the selected token is $5 or less.

(d) The significance level $\alpha = 2/25 = 0.08$ which is less than 0.10.

(e) The chance of a Type II error is $\beta = 11/25 = 0.44$.

(f) Our decision is to reject H_0.

1.23

(a) False.

(b) False.

(c) True.

1.25

(a) The frequency plots are provided below:

```
            Bag X                                          Bag Y
             X                              X                                    X
             X                              X                                    X
      X      X      X                       X                                    X
      X      X      X                       X                                    X
      X      X      X                       X                                    X
      X      X      X                       X                                    X
      X      X      X                       X      X              X              X
      X      X      X                       X      X              X              X
X     X      X      X      X                X      X      X       X              X
_____          _____
Blue  Brown Yellow Green  Red              Blue   Brown  Yellow   Green         Red
```

(b) No, the response being recorded is the color, which has no particular ordering for the outcomes. If the colors had been listed in the order of Blue, Red, Brown, Green, and Yellow, then the frequency plots would be as shown below and the apparent direction of extreme would now be one-sided to the left. So it is not appropriate to discuss a direction of extreme in this case.

1.27

The p-value should be small in order to reject the null hypothesis H_0. A small p-value indicates that the observed data or data even more extreme is very unlikely or unusual if the null hypothesis is true. In general, we reject H_0 if p-value is less than or equal to α, the significance level.

1.29
(a) Frequency plots for the two completing hypotheses.

H_0:

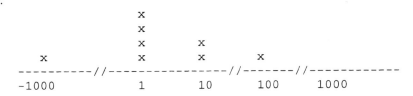

H_1:

```
                                 x
                                 x
                                 x          x
                    x            x          x
         x          x            x          x
-----------//----------------------//--------//------------
    -1000            1           10         100        1000
```

(b) α = chance of a Type I error = chance of rejecting H_0 when H_0 is true
= chance of observing \$1 from the winning bag
= 2/10 = 0.20

(c) β = chance of a Type II error = chance of accepting H_0 when H_1 is true
= chance of observing \$10 or \$100 from the losing bag
= 3/8 =0.375

(d) No, we did not actually observe a voucher.

1.31
(a) The direction of extreme is one-sided to the left (to the smaller values).
(b) The chance of a Type I error is $\alpha = 1/6 = 0.1667$ (found under the Die A model for 1 or less). The chance of a Type II error is $\beta = 7/10 = 0.70$ (found under the Die B model for more than 2).
(c) The p-value is the chance of getting the observed value of 2 or less, assuming the die is Die A = 2/6 = 0.333. Since this p-value is greater than the significance level α of 0.1667, we cannot reject H_0 and conclude that selected die appears to be Die A.

1.33
(a) H_0: The shown bag is Bag A. H_1: The shown bag is Bag B.
(b) The direction of extreme is two-sided.
(c) i.   The p-value is $\frac{4}{40} = 0.10$.
    ii. Yes, since the p-value is $\leq \alpha$.
    iii. No, since the p-value is $> \alpha$.
(d) i.   The p-value is 1.
    ii. No, since the p-value is $> \alpha$.
    iii. No, since the p-value is $> \alpha$.

1.35
(a) All new model 100-watt light bulbs produced at Claude's plant.
(b) H_0: The population of all new model 100-watt light bulbs (produced at Claude 's plant) has an average lifetime equal to 40 hours. H_1: The population of all new model 100-watt light bulbs (produced at Claude 's plant) has an average lifetime greater than 40 hours.
(c) 10%
(d) The p-value can be any value between 0 and 0.10.
(e) Yes, if the p-value is less than or equal to 0.10, then the p-value is also less than or equal to 0.15 so it is significant at the 0.15 level. However, if we only know that the p-value is less than or equal to 0.10, we cannot be sure whether the p-value is also less than or equal to 0.05. Without further information about the value of the p-value we cannot determine if the data would also be significant at the 0.05 level.

1.37

(a) H_0

(b) p-value > 0.10

(c) We accepted H_0, so we could have made a Type II error.

(d) One-sided to the right. We want to see if the numbers have increased.

1.39

(a) We rejected H_0, so we could have made a Type I error.

(b) We decide that the average cost is higher than $350 while it is really not. Maybe you decide that the cost is too high and decide to attend a different college, while in reality you could have attended this college after all.

(c) The p-value is ≤ 0.10.

(d) Yes.

1.41

(a) Possible values are: For study A the p-value is 0.001, for study B the p-value is 0.11, and for study C the p-value is 0.03

(b) Reject H_0 if the p-value is small, so support for H_0 is shown if the p-value is large, the largest p-value is for Study B.

(c) We rejected H_0, but it was true, so a Type I error could have been made.

(d) For Study A: one-sided to the right, Study B: two-sided, Study C: one-sided to the left.

1.43

(a) See the chart below for the alternative hypotheses.

(b) See the chart below for the possible p-values.

(c) The results for Study C had the most support for the null hypothesis since the p-value was the largest.

(d) This would be called a Type I error.

	Null Hypothesis	**Alternative Hypothesis**	***p*-value**
Study A	The true proportion of females is equal to 0.60.	The true proportion of females is **not equal** to 0.60.	**0.08**
Study B	The average time to relief for all Treatment I users is equal to the average time to relief for all Treatment II users.	The average time to relief for all Treatment I users is **less than** the average time to relief for all Treatment II users.	**0.005**
Study C	The true average income of adults who work two jobs is equal to $70,000.	The true average income of adults who work two jobs is **greater than** $70,000.	**0.20**

1.45

(a) The direction of extreme is one-sided to the left.

(b) See graph for shading and labeling of α (the shaded region under H_0) and β (the shaded region under H_1.

(c) See graph for shading and labeling of the p-value (the area to the left of 3.2 under H_0.

(d) Since the p-value is larger than α, the result is not statistically significant.

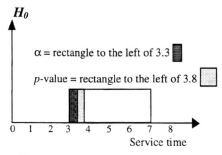

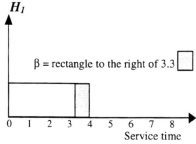

1.47

(a) True.

(b) False.

1.49

(c) to be statistically significant at the 5% level means the p-value is less than or equal to 0.05. However, we do not know if the p-value is less than or equal to 0.01 or if it is between 0.01 and 0.05, so the answer is "sometimes yes" (if the p-value is also $\leq$ 0.01) and "sometimes no" (if the p-value is > 0.01).

1.51

(a) You observed a yellow ball, which is the most extreme result that you could get. With only one observation there is no more extreme than observing a yellow. So the p-value is the chance of observing a yellow ball under the null hypothesis, which is $1/5 = 0.20$. Since the p-value is larger than 10%, the result is not statistically significant.

(b) The data consists of selecting two balls with replacement (and the order is not important). The possible outcomes are shown in the picture below:

We observed one yellow and one blue ball. Results that are even more extreme would be observing two yellow balls. So the p-value is $(1+8)/25 = 0.36$. Since the p-value is larger than 10%, the result is not statistically significant.

1.53

Statement (i) since the effect is small so we would need a larger sample size to detect it.

1.55

(a) (i) α = chance of observing an average ≥ 45 from bag A = $4+2+1/190 = 0.0368$
   (ii) 45
   (iii) $\beta = 0.289$
   (iv) The chance that we decide that the bag shown is bag A, while it really is bag B is 0.28 (or 28%).

(b) (i) First note that the observed average voucher value is $35, so we have:
   p-value= chance of observing an average of $35 or more extreme under H_0 = $17+9+4+2+1/190 = 0.1737$
   (ii) p-value $\leq \alpha$ so we reject H_0
   (iii) $20+17+9+4+2+1/190 = 0.2789$
   (iv) p-value $> \alpha$ so we accept H_0
   (v) The cut-off value is the average of $35, since we rejected H_0 for $35 and we accepted H_0 for $30.

1.57

(a) 20 nCr 2 = 190
(b) 20 nCr 3 = 1140

2.1
35% is a parameter - it is a numerical summary of the population
28% is a statistic - it is a numerical summary of a sample from the population.

2.3
Answer is (c), statistics is to a sample.

2.5
(a) Using the calculator with a seed value of 291, the selected persons are 39 (79 years old), person 3 (75 years old) and person 24 (70 years old). The average age is 74.67 years.
(b) 22 (36 years old), 34 (89 years old) and 29 (89 years old). The average age is 71.33 years, different from the mean calculated in part (a).
(c) 69.19 years.
(d) 74.67 and 71.33 are statistics, while the mean in part (c) is a parameter.

2.7
(a) The population consists of (the planned vote for) the 100 U.S. Senators.
(b) $N = 100$.
(c) The sample consists of (the planned vote for) the ten selected U.S. Senators.
(d) $n = 10$.

2.9
(a) The value of 34 is a statistic.
(b) Response bias.

2.11
Response bias.

2.13
Non-response bias.

2.15
This survey may be subject to nonresponse bias. Only those alumni who respond and report their income will be included. Alumni who perhaps are currently unemployed or in a low paying position may elect not to respond. Therefore, the reported average income based on such a survey may be biased upwards -- the average may be larger than the actual average for all alumni.

2.17
(a) Yes, each sample of size 20 has the same chance as any other sample of size 20 to be selected (assuming all tags are exactly the same and the box is thoroughly mixed).
(b) It is drawn without replacement.

2.19
(a) With the calculator the selected ID numbers are: 179, 2274, 3327. With the random number table the selected ID numbers are: 2398, 2258, 3540.
(b) A statistic.

2.21

(a) H_0: The proportion of dissatisfied customers equals 0.10.
  H_1: The proportion of dissatisfied customers is less than 0.10.

(b) With the calculator the selected costumers are: 34318, 15553, 8461, 614.
  With the random table the selected costumers are: 15409, 23336, 29490, 43127.

(c) Type II error.

(d) 0.21.

(e) No, the p-value is > 0.05.

(f) ii. Statistic.

2.23

(a) Based on her decision rule, Jane rejects H_0.

(b) Since Jane rejected the null hypothesis, the data are statistically significant.

(c) No, based on a sample of two $1 Jane can not be certain of which purse she has since both purses contain at least two $1.

(d) Type I error: reject H_0 when H_0 is true.

(e) Simple random samples of size 2 from null purse:
  1_1 and 1_2, 1_1 and 5_1, 1_1 and 5_2
  1_2 and 5_1, 1_2 and 5_2, 5_1 and 5_2

(f) Simple random samples of size 2 from alternative purse:
  1_1 and 1_2, 1_1 and 1_3, 1_1 and 1_4
  1_2 and 1_3, 1_2 and 1_4, 1_3 and 1_4

(g) The p-value is the chance of getting two $1 bills or more extreme (in the direction of H_0, but in this case, there is no "more extreme") if H_0 is true. The p-value is the chance of getting two $1 bills if H_0 is true, i.e., 1/6.

2.25

(a) 100/200 = 1/2 = 0.50 or 50%.

(b) 100/1000 = 1/10 = 0.10 or 10%.

(c) The chance of being chosen is 0.10. This is NOT a simple random sample. For this stratified random sampling plan each possible sample would contain exactly 100 males and 20 females. All samples of size 120 are not equally likely (as it should be for simple random sample). Some samples of size 120 are not even possible, for example, having 120 males.

2.27

Stratified random sampling. You are dividing up your population by gender then selecting your sample within the strata at random.

2.29

(a) Stratified random sampling.

(b) No, if there are more people whose family name begins with say A and fewer people whose family name begins with Z, then those with Z will have a higher chance of being selected.

(c) Selection bias -- a systematic tendency to exclude those with unlisted phone numbers.

2.31

$(0.20)(16) + (0.50)(43) + (0.30)(71) = 46$ years.

2.33

(a) All former university graduate students.

(b) Stratified random sampling.

(c) False.

2.35

(a) Stratified random sampling.

(b) Selection bias.

2.37

(a) Stratified random sampling

(b) The chance is 0, only 1 of the two fiction books (A, B) will be selected.

(c) Two books that could be selected are Book A and Book C.

(d) Another pair of books that could be selected are Book A and Book D.

(e) The chance that the total number of pages exceeds 800 is the same as the chance that the two selected books are Book B and Book D. There are 4 possible pairs of books that could be selected of which the (B, D) pair is 1, so the chance is $1/4 = 0.25$.

2.39

(a) Stratified random sampling.

(b) High: 20 clients, Moderate: 125 clients, Low: 45 clients.

(c) With the calculator the selected clients are: 163, 2196, 214, 2462, 740.
With the random number table the selected clients are: 1887, 1209, 2294, 954, 1869.

2.41

(a) With the calculator, the label of the first student selected is 3. With the random number table, the label of the first student selected is 3.

(b) The students in the sample are those with ID numbers 3, (3 + 4 =) 7, (7 + 4 =) 11, and (11 + 4=) 15, for a total of 4 students.

2.43

(a) Systematic sampling (1 in 30). Once you select the sample you can determine the number of freshmen you selected. Also knowing the population list you can calculate the number of freshmen you would select for each of the 30 possible samples (depending on the starting point.)

(b) Stratified random sampling. Yes, you would have 25 freshmen.

2.45

(a) i. $0.02(500)+0.03(1200)+0.05(18000) = 10+36+900 = 946$.

ii.   Stratified Random Sampling.

iii. With the calculator the selected labels are 432, 232, 304, 412, 372. With the random number table, we might assign the first High category driver the labels 001 and 501. We would assign the second High category driver 002 and 502. This assignment pattern would continue until the 500th High category driver who would be assigned the labels 000 and 500. Reading off labels from row 60, column 1, we have: 789, 191, 947, 423, and 632. This would correspond to selecting the 289th, 191st, 447th, 423rd, and 532nd High category drivers in the list of 500 High category drivers.

(b) i.   With the calculator or the random number table, the first selected label is 15. Thus the selected labels will be 15, 35, 55, 75, 95, and so on.

ii.   Since the 500 High category drivers divide evenly into groups of 20 (500/20 = 25), there will be a total of 25 High category drivers in the systematic 1-in-20 sample

2.47

False, the chance depends on the number of clusters.

2.49

(a) This is cluster sampling since the students are first divided into clusters (undergraduate classes). A class/cluster is then selected using a simple random sample and all students from that class/cluster are sampled.

(b) No, since not all students take the same number of classes. Students who attend more classes have a greater chance of being selected.

(c) No it is not biased since the cluster was selected at random. It is a case of poor design together with bad luck. When clustering the variability between clusters should not be more important than the variability within clusters. Here we have poor design because there might be more variability between a class with many students on athletic scholarship and a class without any, than variability within each of those classes.

2.51

(a) The type of sampling performed in each dorm is cluster sampling, with the rooms forming the clusters and 3 clusters were selected at random.

(b) If each cluster selected has one student, we would have 3 students from each of the four dorms for a minimum sample size of 12 students.

(c) If each cluster selected has three students, we would have 9 students from each of the four dorms for a maximum sample size of 36 students.

(d) We do not know, it will depend on how many clusters selected are rooms with women. The number of women could be as low as 0 and as high as 36.

(e) No, there will be anywhere from 3 to 9 freshmen students sampled from the freshmen dorm, as well as from 3 to 9 sophomores, from 3 and 9 juniors, and from 3 to 9 seniors.

2.53

(a) It is a systematic 1 in 45 sampling resulting in 50 students (1 from each of the 50 sections, the 33rd in each of the 50 lists of 45 students).

(b) It is a cluster sample and you cannot know the sample size (number of students selected) because we do not know how many students are in the various major clusters.

2.55

(a) False.

(b) False.

(c) True.

2.57

(a) i. Convenience sampling.
  ii. Yes, a selection bias.
  iii. The calculated average is expected to be higher than the true average, as all of the books in the sample have already been checked out at least once, and may include some of the more popular books.

(b) Cluster sampling.

(c) i. Stratified random sampling.

  ii. Overall estimate: $\left(\dfrac{400}{1200}\right)(20) + \left(\dfrac{200}{1200}\right)(15) + \left(\dfrac{600}{1200}\right)(10) = 14.2$ times checked out.

(d) For each of the three categories of books the following stages are followed.
Stage 1: Divide the books into clusters according to the last digit of the call number (0 through 9). Take a simple random sample of 3 digits from the list of 0, 1, 2, 3, 4, 5, 6, 7, 8, and 9.
The clusters of books (in that category) with call numbers ending with those selected digits are selected.
Stage 2: Within each of the selected clusters of books from Stage 2, select a simple random sample of 7 books.
Note that with this multistage sampling plan, we will have a total of 3 categories x 3 clusters x 7 books = 63 books.

2.59

(a) With the calculator or the random number table, the selected region is 3 = Southwest.

(b) Stratified random sampling.

(c) i.   1-in-10 systematic sampling.

  ii.   0.10.

  iii. With the calculator, the first five selected cans are 7, (7 + 10 =) 17, (17 + 10 =) 27, (27 + 10 =) 37, and (27 + 10 =) 47. With the random number table we might label the first can 1, the second can 2, ..., and the 10th can 0. Then the first five selected cans are 7, (7 + 10 =) 17, (17 + 10 =) 27, (27 + 10 =) 37, and (27 + 10 =) 47.

  iv. $\dfrac{125}{10} = 12\dfrac{5}{10} \Rightarrow$ 12 or 13 cans. However, there will not be a 7th can to select in last group. Thus the total number of cans in the sample will be 12.

(d) i. H_0

  ii   Two possible values are 0.12 and 0.15.

  iii. Yes, a Type II error.

Interactive Statistics 2nd Edition: Chapter 3 Odd Solutions

3.1
(a) Explanatory variable: Amount of coffee consumed.
Response variable: Exam performance.
(b) Explanatory variable: Hours of counseling per week.
Response variable: Grade point average.
(c) Explanatory variable: Type of driver (levels are good or bad).
Response variable: Reaction time on a driving test.

3.3
(a) Observational study.
(b) Blood cholesterol level.
(c) Frequency of egg consumption.
(d) Diet, age, amount of exercise, and history of high cholesterol are a few possible confounding variables.

3.5
(a) Observational study.
(b) Homelessness.
(c) Some are: separation status (whether or not they were separated from their parents), poverty Status (whether or not they were raised in poverty), family problem status (whether or not they have family problems, abuse status (whether or not they were sexually or physically abused).
(d) All homeless people in Los Angeles.
(e) The homeless people in Los Angeles that were surveyed.
(f) Statistic.

3.7
(a) H_1.
(b) The p-value was less or equal to 0.05.
(c) iii. Observational study, Prospective.

3.9
(a) Observational study (retrospective).
(b) Response variable is kidney stones status (whether or not the subject develops kidney stones); Explanatory variable is amount of calcium in diet.
(c) A diet high in calcium lowers the risk of developing kidney stones in men and women.

3.11
(a) Observational study (retrospective).
(b) Response variable is Alzheimer's disease status, explanatory variable is linguistic ability.
(c) The chance of low linguistic ability given a person has Alzheimer's.

3.13
(a) Durability.
(b) Dye color and type of cloth.
(c) 20
(d) 20 x 6 = 120.

3.15

(a) Batches of feed stock.

(b) Yield of the process.

(c) Temperature (2 levels) and Stirring Rate (3 levels).

(d) 6 treatments.

(e) 24 units.

(f) Design Layout Table:

Factor 1: Temperature

		50	70
Factor 2:	60	4 batches	4 batches
Stirring Rate	100	4 batches	4 batches
	140	4 batches	4 batches

3.17

(a) Weight.

(b) Fiber content and carbohydrate content.

(c) Fiber levels (3) Low, Medium and High. Carbohydrate levels (2) Low and Medium.

(d) $3 \times 2 \times 20 = 120$.

3.19

The rats are labeled 1 through 8. Use a calculator with seed= 1209 or a random number table with row=24, column=6 to select the 4 rats to receive the treatment. The other 4 rats will not receive the treatment. At the end of the week the effectiveness of the vaccine against the virus will be measured.

3.21

(a) Experiment.

(b) Cause of death.

(c) Estrogen therapy with two levels: receive the treatment, do not receive the treatment.

(d) Placebo.

H_0: Women who receive Estrogen therapy do not have a lower risk of dying from a heart disease than women who do not receive the Estrogen therapy

H_1: Women who receive Estrogen therapy have a lower risk of dying from a heart disease than women who do not receive the Estrogen therapy.

(e) p-value was less or equal 0.05.

(f) Yes, Type I error.

3.23

Answer is (d).

3.25

(a) With the calculator the selected rats were: 14, 13, 5, 16, 9, 3, 4, 2, 7, 10. With the random number table the selected rats were: 4, 13, 8, 5, 12, 15, 2, 6, 9, 20.

(b) Have another researcher make and record the measurements.

3.27

Observational study. No active treatment was imposed.

3.29

(a) Observational study.
(b) Attitude towards mathematics.
(c) Gender.
(d) Cluster sampling.
(e) Selection bias.

3.31

(a) It is not clear from the article.
(b) It may not be appropriate to extend these results to the general male population, nor to the female population.
(c) Confounding variables.

3.33

(a) Temperature (3 levels) and Baking Time (2 levels).
(b) Taste.
(c) 6 treatments.
(d) 36 batches.

3.35

Lack of blinding of the subjects and of the experimenter(s) or evaluator(s).

3.37

(a) iii. confounding variable.
(b) i.   False
    ii. False
    iii. True
    iv. False

3.39

The placebo effect is a phenomenon in which receiving medical attention, even administration of an inert drug, improves the condition of the subjects.

3.41

(a) Experiment.
(b) Proportion of juice added to the drink.
(c) Rating of the juice in on a scale of 1 to 100.
(d) iii. Confounding variable.
(e) ii.   Single-blinded.
(f) With the calculator, all five volunteers received 5% juice.
    With the random number table, the first four volunteers received a 10% juice and the last one received 5% juice (using 0-4 to represent 5%, using 5-9 to represent 10%).
(g) False.

3.43

(a) Experiment.
(b) Intestinal discomfort measured on a scale (0-5).
(c) Explanatory variable is type of milk. The treatments are treated milk (has a lactose enzyme) and ordinary milk.
(d) The population under study is people who say they are lactose intolerant. The sample is not random because they are volunteers.

3.45

(a) Observational study.
(b) We don't know whether the patients in both groups were in the same physical condition.
(c) Assign to the 100 patients numbers 1-100. With the calculator or with the random number table select at random the first 50 patients. Those patients will receive treatment 1, the other patients will receive treatment 2.
(d) No. The treatment is an operation.

3.47

(a) Experiment.

(b) Using the calculator . With seed = 23 the selected men were: BJ(3), AB(14), JP(4), ZB(16), TN(20), CF(17), JW(7), TD(19), MK(2), SK(5).

Using the calculator . With seed= 41 the selected women were: NL(12), BG(2), MM(5), JB(4), AK(10), KS(6), KB(18), SL(11), JG(19), NI(17).

With the random number table. Men(row 12, column1): RM(9), TG(18), MK(2), VN(8), WB(12), SK(5), BH(11), BJ(3), RM(9), JD(1).

With the random number table. Women(row 22, column 1): CI(16), NI(17), MA(3), SM(7), BG(2), KS(6), EG(15), AK(10), MM(5), CJ(13).

(c) Confounding variable.

3.49

(a) Using the calculator with seed= 45, the first five subjects selected are: 3351, 3140, 860, 703, and 5488.

With the table of random numbers (row 10, column 5) the first five subjects selected are: 5368, 5753, 3425, 3988, and 5306.

(b) The placebo effect is a phenomenon in which receiving medical attention, even administration of an inert drug, improves the condition of the subjects.

(c) 3.5% of 4058 = 142.

(d) Statistic.

(e) H_0

(f) p-value > 0.05.

(g) Type II error.

3.51

(a) Observational study.

(b) Explanatory variable: Amount of wine.

Response variable: IQ score.

(c) Amount of wine was likely self-reported and may result in a response bias.

(d) Who paid for the study? Who were the subjects on the study? How many subjects were selected?

3.53

(a) Experiment.

(b) Explanatory variable: Behavioral therapy status.

Response variable: Insomnia status.

(c) The number of people in each treatment group is unknown. We only know that they were 75 in total.

(d) How were the results analyzed?

How were the 75 subjects selected? Was the placebo treatment truly a placebo?

3.55

Answers will vary.

4.1
(a) qualitative
(b) quantitative-continuous
(c) qualitative
(d) quantitative-discrete
(e) qualitative
(f) quantitative-continuous

4.3
(iv) Qualitative

4.5
The total number of students who failed the midterm may be a valid measure of the difficulty of the class. In addition to that information, one needs information regarding the total students in each class. There may be only 20 students in the English class but a total of 200 students in the Chemistry class. A more useful measure may be the proportion or percent of students who failed the midterm. Other questions that one may ask: What type of students takes each class? Is the class a requirement?

4.7

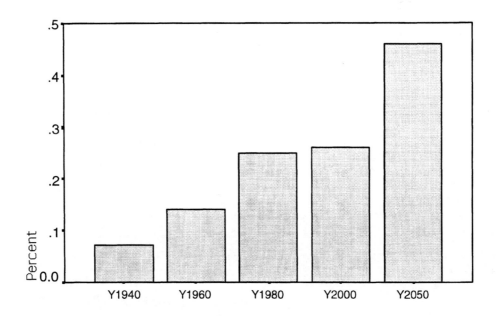

4.9

(a) The sales are represented by the areas of the bars, not just the height of the bars.

(b)

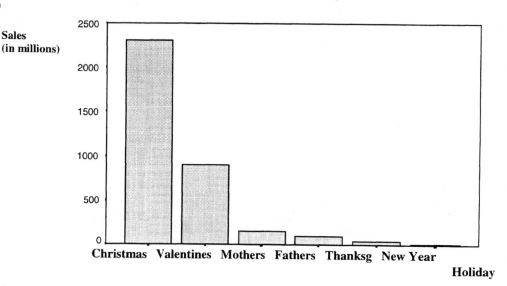

4.11

(a) $0.20(620) = 124$

(b) No, background is a categorical variable. It is not appropriate to discuss shape in a bar graph since the order of the categories is arbitrary.

4.13

(a) The conditional distribution of years of experience given gender is:

		Years of Experience		
		1-6	**7-12**	**13+**
Gender	*Male*	25.7%	37.9%	36.4%
	Female	42.3%	35.1%	22.6%

(b) It appears that male coaches generally have more years of coaching experience than females. However, there have been more women (in terms of percentages) entering the field of coaching. Since Title IX went into effect, and the interest in women's sports is rising, many new coaching jobs have been available for women's sports. So a high percent of relatively *new* female coaches is reasonable.

4.15

(a) Men: $16,120/(16,120+1,728+1,708+3,751+1,462) = 16,120/24,769 => 65.08\%$
Women: $2,406/(2,406+670+522+810+1633) = 2,406/6,041 => 39.83\%$

(b) The conditional distribution is:

		Method of Suicide				
		Firearms	*Drugs*	*Hanging*	*Gases*	*Other*
Gender	*Men*	65.1%	5.9%	15.1%	6.9%	7.0%
	Women	39.8%	27.0%	13.4%	8.6%	11.1%

(c) The bar chart is shown below:

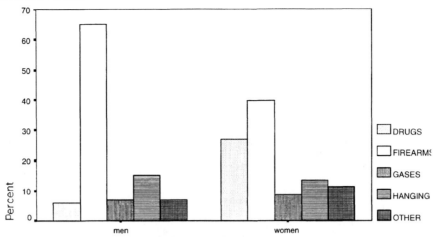

(d) Men who commit suicide overwhelmingly choose firearms while women are more evenly split between firearms and drugs.

4.17

(a) 36/50 = 0.72 or 72%

(b) 24/50 = 0.48 or 48%

(c) Yes, since the p-value for the test of no association is 0.01431, which is less than the 0.05 level of significance.

4.19

Frequency Plot for the number of goals scored during each game last season.

```
                X
        X       X       X
        X       X       X
X       X       X       X       X       X
X       X       X       X       X       X                       X
────────────────────────────────────────────────────────────────
0       1       2       3       4       5       6       7       8       goals scored
```

4.21

Answers will vary.

4.23

Stem-and-leaf plot: Note: 83 | 4 represents an octane rating of 83.4.

```
83 | 4
84 |
85 |
86 | 7
87 | 5 7
88 | 0 4 5 8
89 | 1
90 | 5
```

4.25

(a) Back-to-back stem-and-leaf plot. Note: 14 | 5 represents a score of 45 points.

```
      Student B              Student A
                      14 | 5 8
                      15 | 2
                      16 | 9
          8 6 5 4 3 | 17 | 6 9
      7 4 2 1 0 | 18 | 0 6
                      19 | 2 4
```

(b) Student A's scores range from 45 to 94. Student A did receive the highest score of the two students. However, Student A also received the lowest score of the two students. The scores for Student B ranged from 73 to 87. Student B performed very consistently in the 70's and 80's.

4.27

(a) 781 hours (a Longlife light bulb).
(b) If you wanted the brand that gives the longest lasting light bulb.
(c) If you wanted the brand that lasts longest on average (that is using the mean).

4.29

(a) Note: 80|8 represents an On-Time departure rate of 80.8%.

% On-Time Arrival		% On-Time Departures
64	75	
0	76	
8100	77	
	78	
6	79	
1	80	8
5211	81	
5	82	
1	83	1389
91	84	3
8	85	77
40	86	156
9	87	06
	88	1689
	89	089
3	90	
	91	3
	92	6

(b) The arrivals range from 75.4% to 90.3%. The arrival values are quite dispersed with a gap on the high side between Detroit (87.9%) and Raleigh (90.3%). The departures range from 80.8% to 92.6%. There is less variation among the departures as compared to the arrivals. The departures distribution is somewhat bimodal with a gap on the low side between Chicago (80.8%) and Denver (83.1%).
(c) Airports perform better overall with respect to on-time departures. The center value for on-time departures (86.8%) is higher than the center value for on-time arrivals (81.15%). The range for the on-time departures is smaller than for the on-time arrivals. With the exception of Chicago, all of the % on-time departures are higher than the center value for the % on-time arrivals.

4.31

(a) 3 + 6 + 8 + 10 + 5 = 32.
(b) 9/32 = 0.2813

4.33
(a) The distribution is skewed to the right.
(b) 5/31 => about 16%
(c) Since 5 minutes corresponds to 300 seconds, we have 2 people.
(d) We cannot state the maximum length of time a customer had to spend in line. All we know is that a customer spent between 350 and 400 seconds in line but we don't know the actual time.

4.35
Back-to-Back Histograms

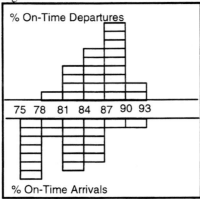

4.37
(a) For the end of 1925, the shape is somewhat uniform but with an extreme modal class of 100+. For the end of 1960, the shape is positively skewed (or skewed to the right), with the most frequent price class of 20-25. For the end of 1994, the shape is skewed to the right with the most frequent price class of 10-15.
(b) The general shift has been from many stocks priced 100+ to more stocks priced between 5-40.
(c) Yes. A lot of stock splits would lead to more stocks at lower prices.

4.39
Time Series I matches Description (1) since we would expect the proportion of babies born that are girls to fluctuate around 0.50. Time Series II matches Description (2) since cumulating the number of batteries that fail would increase over time.

4.41
(a) Time plot.

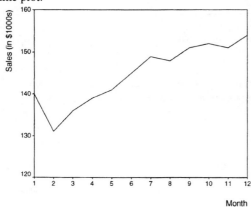

(b) an increasing trend.

4.43

(a) 25% National, 30% Michigan

(b) 18% Michigan, 17% National

(c) That information is not given. It is not true that since 25% of the fatal accidents in Michigan involved female drivers, then 75% of the fatal accidents involved male drivers because a male and female driver may both die in a crash.

(d) In 1983, both the Michigan and the National percentages were around 20%. For Michigan that was a decrease from the previous year, however, nationwide, this was a slight increase from 1982. Both for the Michigan and the National numbers, the 1984 numbers were up from 1983. In Michigan there was a slight dip in the numbers in 1983, nationwide there was not.

(e) More women are working; consequently, more women are driving.

4.45

(a) Plot C

(b) Negative linear association.

4.47

Let X be the independent variable and Y be the dependent variable

(i) (a) X = Year; Y = Population of the U.S.

(b)

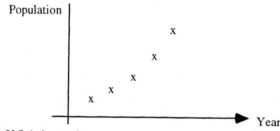

(c) The population of the U.S. is increasing

(ii) (a) X = Time; Y = Temperature of a drink

(b)

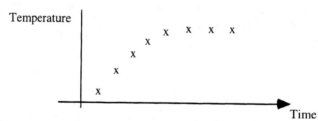

(c) The temperature increases at first and after a while levels off to approximately room temperature.

4.49

Let X = hours of exercise per week, and Y = level of cholesterol. As you increase the number of hours of exercise, (hopefully) your level of cholesterol will decrease. Another example is: Let Y = the temperature of a cup of hot tea outside the house in the winter; and X = the length of time it is outside the house.

4.51

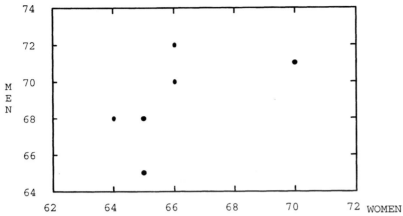

The scatterplot shows a positive association between male height and female height. The linearity might be reasonable but it is not a very strong linear association.

4.53

(a)

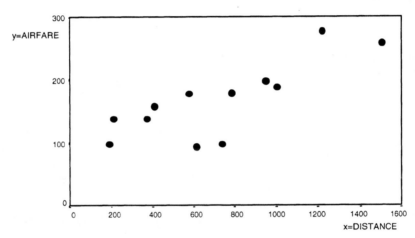

(b) Yes, positive.
(c) Fairly strong.

4.55

(a) Median income decreased to nearly $36,000.
(b) Median income increased from $35,000 to nearly $40,000.
(c) Median income decreased back to approximately $37,000.
(d) Predications would be difficult without further information to perhaps help explain the changes described in parts (a), (b), and (c).

4.57

(a) You might ask: Is this an average SAT score? What data were used to get this average? Is the spending an average amount spent? Across all schools? Just some schools? Both large and small schools?
(b) The actual SAT score range is: 400 - 1600. With the updated graph the increase in spending and decrease in SAT would be less dramatic.

4.59
(a) quantitative continuous
(b) quantitative continuous
(c) quantitative continuous
(d) quantitative discrete
(e) qualitative
(f) quantitative discrete
(g) quantitative continuous

4.61
(a) We have $4179+3766+251 = 8196$ students.
(b) 88 of the 251 students, or $88/251 => 35.06\%$.
(c) The pie chart would look the same with 35.06% for Hispanic, 35.05% for Black, 17.1% for White, and 6.8% for Asian. You might also include an "Other" category with the remaining 6%.

4.63
Answer is: (d) The conditional distribution of Blood Type given Rh Factor.

4.65
(a) Observational study.
(b) Color blindness (with responses being yes or no).
(c) Gender (with levels being male or female).
(d) Statistic (2% is a sample percentage).
(e) Stratified with female = stratum I and male = stratum II.
(f) (i) Yes; (ii) Can't tell; (iii) No
(g) (ii)

4.67
(a) Time is continuous, but since it was recorded to the nearest 5 minute interval it is discrete.
(b) The distribution is skewed to the right with an outlier at 105 minutes. The waiting times range from 5 minutes to 105 minutes. The median is 32.5 minutes.
(c) Since the distribution of waiting times is skewed, the mean waiting time is not an appropriate measure of the typical waiting time. The median is resistant to outlier and skewed distributions. Therefore, the median (32.5 minutes) is an appropriate typical waiting.
(d) It does not say how that day was selected or whether the 30 patients were all of the patients on that day or a sample.

4.69
(a) Amount of Radiation Emitted (mR/hr).
(b) No. We only know that the range falls between 0.1 and 0.9. We don't know the exact upper and lower limits of the range.
(c) There are 10 out of 20 stores for 50%.
(d) Since $0.5/0.08 = 6.25$, the maximum we could use is 6 small television sets.

4.71

(a) Time plot.

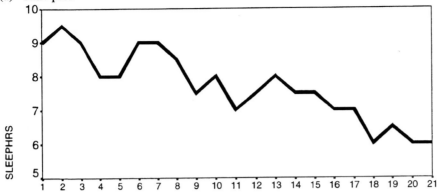

DAY

(b) (ii) decreasing trend.

4.73

(a) Time plot.

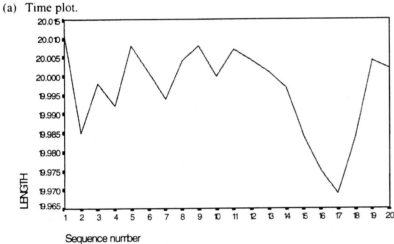

Sequence number

(b) Observations 1 through 13 seem to be varying around the target value of 20.000 cm. Observations #16 and #17 are extremely low.

(c) What happened after observation #13? Did the operator fall asleep? Did the settings get bumped?

4.75

(a) Scatterplot.

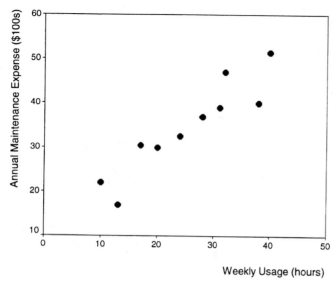

(b) Yes, the relationship does appear to be linear.
(c) The relationship appears to be positive.
(d) There are no unusual values or outliers.

4.77

(a) A person born in 1946 would be 19 years old in 1965 and a person born in 1964 would be 1 year old (or younger) in 1965.
(b) Some of the 1,867,000 people being tracked over time have died.
(c) No, there heights of the bars are not given, nor any axis with values to be able to approximate the heights.
(d) The distribution is skewed to the right for both 1965 and 1995, with the skewness being stronger in 1965. In projected 2025 the distribution is nearly uniform.

4.79

(a) The enforcement officers in Utah were assaulted more often between 10 PM and 2 AM and less often between 6 AM and 8 AM.
(b) The second figure shows the continuity of time and better depicts the most common times for assaults to occur.
(c) How was "assault" defined? Were these Utah data based on all cities, counties, rural areas?

5.1
(c) mean

5.3
The overall average for the 40 students would be: [25(82) + 15(74)]/40 = 79.

5.5
The median may have been preferred if the distribution of resale house prices was skewed to the right. The median is a more resistant measure of center, and would not be so greatly affected by the few very expensive homes.

5.7
(a) Known: $\dfrac{X_1 + X_2 + X_3 + X_4 + X_5 + X_6 + X_7 + X_8 + X_9 + X_{10}}{10} = 35$,

thus, $X_1 + X_2 + X_3 + X_4 + X_5 + X_6 + X_7 + X_8 + X_9 + X_{10} = 350$

now enter $X_{11} = 32$, the new mean is :

$\dfrac{X_1 + X_2 + X_3 + X_4 + X_5 + X_6 + X_7 + X_8 + X_9 + X_{10} + X_{11}}{11} = \dfrac{350 + 32}{11} = 34.727$

(b) We cannot find the new median because we do not know where the new age will fall in the order of ages.

5.9
(a) Yes, I agree. The mode is the value occurring most frequently.
(b) For the values: 1, 1, 1, 2, 20; the median = 1 = minimum and the mean = 5. So 4 out of 5 or 80% of the values are below the mean

5.11
(a) (i)   s = 9.016
     (ii) s = 0
     (iii) s = 2.669
(b) The standard deviation is so large due to one very large outlier, namely the value of 27.
(c) (i)   range = 27 – 1 = 26
     (ii) range = 7 – 7 = 0
     (iii) range = 11 – 3 = 8
     The range is misleading for data set (i). If the outlier of 27 were removed, the range would be 8.

5.13
(a) From the box plot, the highest start-up cost is about $670,000.
(b) From the histogram, the distribution appears to be skewed to the right.
(c) The mean, since the box plot shows that the median is less than $100,000.
(d) It is the largest value that is not considered a potential outlier, the largest observation that falls within the upper inner fence.

5.15
(a) True
(b) True
(c) False
(d) True
(e) False

5.17
Part (b)

5.19

(a) Min = 28, Q1 = 32, Median = 41, Q3 = 48, Max = 50.5
(b) 15 years old
(c) 31 years old
(d) IQR = 48 – 32 = 16 years
(e) 1.5(16) = 24; lower fence = 32 – 24 = 8; Since 20 is > 8, the observation would not be an outlier.

5.21

(a) equal to
(b) larger than
(c) Larger than, because the mean will be in the middle so there are many large deviations. Whereas for plot 2, the mean is drawn to the left where most of the observations are, and for those observations the deviations are small.
(d) Plot 1: range = 4, mean = 2, standard deviation = 1.46; Plot 2: range = 4, mean = 1.33, standard deviation = 1.29.

5.23

The mean in Celsius would be found as: (5/9)[28-32] = -2.22, the standard deviation in Celsius would be found as: |5/9|(10) = 5.55.

5.25

(a) On average the temperatures were about 3.43 °F from their mean of 953.2 °F.
(b) Min = 949, Q1 = 950, Median = 953.5, Q3 = 955, Max = 959
(c) It could increase as much as you want. It will never change the median, since it will still be the middle value.
(d) We can see that there is an upward trend, the temperature seems to increase slowly.

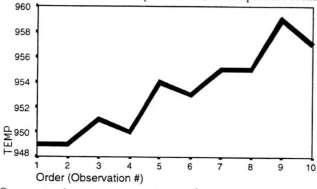

(e) On average the temps were about 1.91 °C from their mean of 511.78 °C.
(f) Min = 509.44, Q1 = 510, Median = 511.94, Q3 = 512.78, Max = 515

5.27

(a) Quantitative Standard Score = -0.41
   Analytical Standard Score = 0.70
   Verbal Standard Score = 1.91
(b) Verbal Comprehension

5.29

We have that [(70)(52)+(30)(lab score 2)]/100 = 64. So lab score 2 = [(100)(64) - (70)(52)]/30 = 92.

5.31

(b) The mean would be higher than the median.

5.33

Yes. For the values 1, 3, 20, the mean = 8 and the standard deviation = 10.44.

5.35

(a) These data appear to be skewed to the left with an outlier at zero.

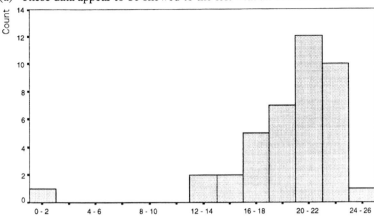

Midterm Scores

(b) The five-number summary would be preferable since the extreme outlier of zero would influence the mean. Min = 0, Q1 = 17.5, Median = 20, Q3 = 22, Max = 24.

(c) With the outlier removed, the mean and standard deviation may now be appropriate measures to summarize these data. Mean = 19.564, Standard deviation = 2.927.

5.37

(a) Negative means that the after is larger than the before; blood pressure increased.

(b) Sample mean = 5; sample standard deviation = 8.74.

5.39

(a) The best performance by the old model resulted in a 60 bottles stuffed in a minute.

(b) Approximately 25% of the time the old model stuffed at least 52 bottles per minute.

(c) On average, the number of bottles stuffed per minute for the new model was 64.05, give or take about 3.15.

(d) Min = 58, Q1 = 62, Median = 64, Q3 = 66.5, Max = 70

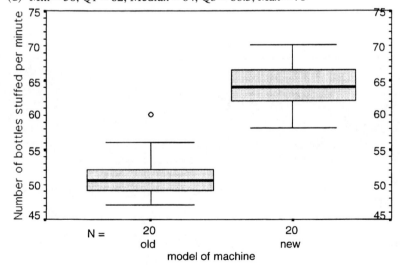

5.41

(a) False: Q1 and the median of 1, 2, 2, 2, 2, 2, 3, 4, 5 are both 2, whereas Q3 is 3.5.

(b) False: If there are 90 boxes that weigh 4 lbs, and 10 boxes that weigh 14 lbs, the total average is [(90)(4)+(10)(14)]/100 = 5 lbs, but there are 90 boxes that weigh less than 5 lbs and only 10 that weigh more than 5 lbs.

(c) False: Both of the following data sets have a mean of 3 and a standard deviation of 1. However, their histograms are not the same. Data set 1: 1, 3, 3, 3, 3, 3, 3, 3, 5; Data set 2: 2, 3, 4.

5.43

(e) none of the above -- From a symmetric boxplot you cannot conclude whether or not its distribution is symmetric so (a) is false. Although the boxplots are exactly the same, the corresponding distributions may not be the same so (b) is false. Since boxplots give the five-number summary and not the average or mean, (c) is false.

5.45

(a) Yes.

(b) Yes.

(c) Happier if the standard deviation is 4, since then your score is 3 standard deviations above the average, a very high score.

5.47

(a) Five-number summary: min=16.55, Q1=26.95, median=45.585, Q3=51.90, max=69.85

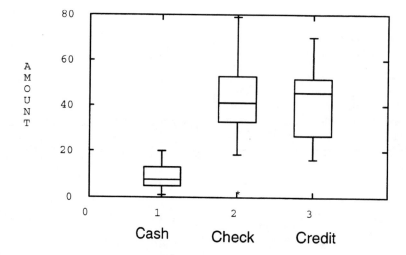

(b) Mean = 41.65, Standard Deviation, s = 14.06

(c) Proportions: cash: 38/100, check: 40/100, credit card: 22/100. About half of the credit card customers came from the customers who paid with cash, and half came from the customers who paid with checks.

(d) From the boxplot, the outlier is at about $1.

(e) Yes, because the cash customers who switch will spend more on average, while check customers who switch will still spend about the same on average.

5.49

The 20 represents 2 standard deviations.

5.51

(a) Time plot.

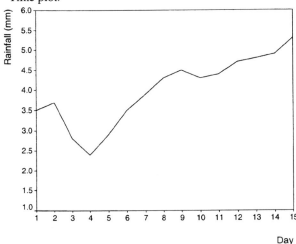

(b) (ii) increasing trend
(c) Min = 2.4 mm, Q1 = 3.5 mm, Median = 4.3 mm, Q3 = 4.7 mm, Max = 5.3 mm.
(d) IQR = 4.7 – 3.5 = 1.2 mm.
(e) (iii) the new median = the old median.
(f) (ii) will be larger than the old mean.

5.53

(a) (ii) a prospective observational study
(b) Min = 14, Q1 = 24, Median = 27.5, Q3 = 32.5, Max = 44.
(c) (i) False
     (ii) Can't tell
     (iii) Can't tell
     (iv) True

5.55

(a) Station 2 with a range of 20 – 7 = 13.
(b) Can't tell.
(c) (i) $\bar{x} = 9.67$, s = 3.14
     (ii) Min = 3, Q1 = 9, Median = 10.5, Q3 = 11, Max = 15
     (iii) IQR = 11 – 9 = 2; 1.5(2) = 3; Lower fence = 9 – 3 = 6; Upper fence = 11 + 3 = 14. So any observation below 6 or above 14 is a possible outlier, namely, 3, 5, and 15.
     (iv) Modified boxplot is provided.

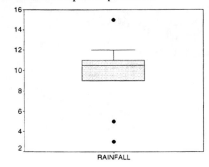

     (v) $\bar{x} = 10.67$, s = 3.14

Interactive Statistics 2ⁿᵈ Edition: Chapter 6 Solutions

6.1
They are all the same.

6.3
(b) The mean of the distribution becomes zero and the standard deviation becomes 1.

6.5
(a) $(680 - 500)/100 = 1.8$.
(b) $(27 - 18)/6 = 1.5$.
(c) Eleanor.

6.7
(a) Prop($Z < 3.49$) = normalcdf (-E99, 3.49,0,1) = 0.9998.
(b) Prop($Z < 4$) = normalcdf(-E99,4,0,1) = 1.
(c) Prop($Z > -0.84$) = normalcdf(-0.84, E99,0,1) = 0.7995.
(d) Prop($Z < 0.84$) = normalcdf(-E99, 0.84,0,1) = 0.7795.
(e) Prop($Z > 2.31$) = normalcdf(2.31,E99,0,1) = 0.0104.
(f) Prop($-3.17 < Z < -1.84$) = normalcdf(-3.17,-1.84,0,1) = 0.0321.
(g) Prop($0.17 < Z < 2.12$) = normalcdf(0.17, 2.12,0,1) = 0.4155.

6.9
(a) Prop($X < 70$) = Prop($Z < (70-63)/5$) = normalcdf(-E99, 70, 63, 5) = 0.9192 or 91.92%.
(b) Prop($X < 70$) = Prop($Z < (70-67)/4$) = normalcdf(-E99, 70, 67, 4) = 0.7734 or 77.34%.

6.11
(a) $\mu = 7$.
(b) 68%.

6.13
(a)

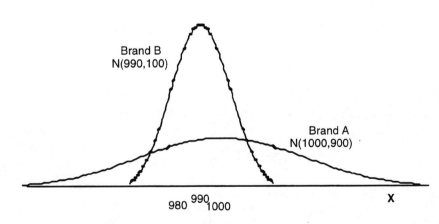

(b) Brand A: 0.7486; Brand B: 0.8413. So a higher proportion of Brand B fuses last longer than 980 days as compared to Brand A.

6.15
Find x, such that Prop($Z < (x-30)/5$) = 0.05. invNorm(0.05,30,5) = 21.78 The maximum duration is 21.78 minutes.

6.17
(a) 0.6915.
(b) 2.8 years.

6.19

(a)

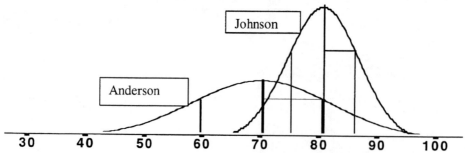

(b) Both give the same. Johnson : Prop(Z > (90-80)/5) = normalcdf(90, E99, 80, 5) = 0.0227.
Anderson : Prop(Z > (90 -70)/10) = normalcdf(90, E99, 70, 10) = 0.0227.

(c) Johnson : Prop(Z < (50 - 80)/5) = normacdf(-E99, 50, 80, 5) = 0.00000000099.
Anderson : Prop(Z < (50 - 80)/5) = normalcdf(-E99,50, 70, 10) = 0.02275.
Anderson gives a higher proportion of E's.

(d) Find x such that Prop(Z < (x - 69)/9)= 0.90; invNorm(.90, 69, 9) = 80.53.

6.21

(a) 0.0047.

(b) 56.7 lesions.

6.23

(a) H_0

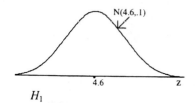

H_1

(b) ii   p-value = Prop(Z > (4.8-4.6)/0.1) = normalcdf(4.8, E99, 4.6, 0.1) = 0.02275.

(c) We cannot reject H_0 because the p-value is > 0.01.

6.25

(a) α = Prop(Z < (8 – 15)/3) = normalcdf(-E99, 8,15,3) = 0.0098.

(b) β =Prop(Z > (8 – 10)/3) = normalcdf(8, E99, 10, 3) = 0.7475.

(c) p-value = Prop(Z < (8,5 – 15)/3) = normalcdf(-1E99, 8.5, 15, 3) = 0.01513

6.27

(a) 1.28 standard deviations below the mean for A/B cutoff and
1.28 standard deviations above the mean for B/C cutoff.

(b) Mean = (50 + 250) /2 = 150   and   (250-150)/ σ = 1.28, so $\sigma \approx$ 78.

6.29

(a) $\bar{x} = 255.4$ grams, $s = 2.769$ grams.

(b)

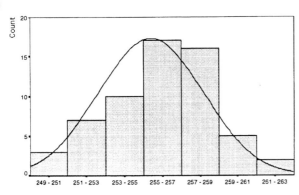

WEIGHT

The distribution appears to be approximately symmetric, bell-shaped, unimodal.

(c) The percentages are 71.7%, 95%, and 100%.

(d) These data appear to resemble a normal distribution approximately.

(e) The product is being filled within the required range approximately 90% of the time.

6.31

(a) The proportion is approximately zero. This could be considered the p-value since it is the chance of the data or more extreme data being observed if the null hypothesis is true.

(b) Proportion $(X < 9)$ = practically zero. We reject H_0 since the p-value is nearly zero.

6.33

(a) 0.67.

(b) 0.33.

6.35

(a)

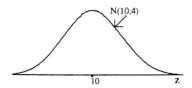

(b) i. False.

  ii. True.

6.37

(a)

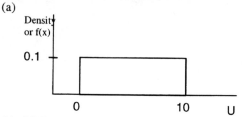

(b) 3/10 or 30%.

(c) i. one-sided to the right.

 ii p-value = 1/10 = 0.10

 iii Reject H_0 since p-value < $\alpha = 0.15$.

6.39

(a)

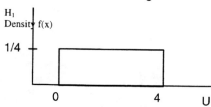

(b) ii $\alpha = (3.2 - 3)/4 = 0.05$

 iv $\beta = (4 - 3.2)/4 = 0.20$

(c) ii. p-value =$(4.1 - 3)/4 = 0.275$.

(d) No, since the p-value = $0.275 > \alpha = 0.05$.

6.41

(a) 25%.

(b) More than 1 pound.

(c) More than 1 pound.

6.43

0.20

6.45

(a)

Y	brown	red	yellow	green	orange	blue
Proportion.	0.30	0.20	0.20	0.10	0.10	0.10

(b) No, because the coding is arbitrary -- if you change the coding, you could change the shape. The variable is qualitative (categorical).
(c) Approximately $1/5 = 0.20$ or 20% should be red.
(d) Take some data (some randomly selected bags). Observe the actual number of each color in the bags. See if the proportions are "close" to those stated.

6.47
Proportion $(X < 36)$ = normalcdf(-E99,36,60,8) = 0.0013.

6.49
(a) Proportion $(64.5 \leq X \leq 72)$ = P((64.5-63.6)/2.5 < Z < (72-63.6)/2.5) = normalcdf(64.5,72,63.6,2.5) = 0.3590.
(b) 1000 (0.359) = 359 women.
(c) 67.7 inches.

6.51
(a) Proportion$(X > 31.5)$ = Prop(Z > (31.5-30)/3) = normalcdf(31.5, E99, 30,3) = 0.3085 or 30.85%.
(b) 34.93 pounds.

6.53
(a) 557
(b) Proportion$(X < 507)$ = 0.6736.

6.55
Mean = 9.75, Standard deviation = 0.375.

6.57
(a) Proportion$(X > 50)$ = Prop(Z > (50-45.5)/1.5) = normalcdf(50,E99,45.5,1.5) = 0.0013.
(b) 0.0015.
(c) 48.58.

6.59
(a) (i) α = normalcdf(13.2, E99, 10, 2) = Prop($X > 13.2$) = Prop(Z > (13.2-10)/2) = 0.055
   (ii) β = normalcdf(-E99, 13.2, 12, 2) = Prop($X < 13.2$) = Prop(Z < (13.2-12)/2) = 0.7257.
(b) p-value = normalcdf(11.2, E99, 10, 2) = Prop($X > 11.2$) = Prop(Z > (11.2-10)/2) = 0.274.

6.61
(a) p-value = Proportion $(X \geq 62.5)$ = Prop(Z$\geq$ (62.5-50)/5) = normalcdf(62.5,E99,50,5) = 0.062.
(b) We reject H_0 at the 10% significance level, since the p-value is < 0.10.

6.63
Proportion$(X > 20)$ = Prop(Z > (20-19)/1.7) = normalcdf(20,E99,19,1.7) = 0.2810.

6.65
(a) Find x such that Prop($X < x$) = 0.95;   invNorm(0.95, 15, 3) = 19.93 minutes.
(b) Mean of Y = 1.2 (15) – 4 = 14 minutes.
   Standard deviation of Y = 1.2 (3) = 3.6 minutes
(c) Find x such that Prop($X < x$) = 0.95;   invNorm(0.95, 14, 3.6) = 19.92 minutes.

6.67

(a)

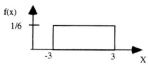

(b) Q1=-1.5 and Q3=1.5, so the IQR=1.5 - (-1.5) = 3. The IQR is not equal to 50%; rather it provides the range of values that cover the middle 50% of the values.

(c)

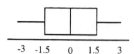

6.69

(a) 0.06.

(b) 0.61.

(c) Skewed to the right.

(d) The median will be between 1 and 2. The mean will be larger than the median.

6.71

(a) Proportion$(X > 2)$ = Prop$(Z > (2-1)/2)$ = normalcdf$(2, E99, 1,2)$ = 0.3085 or 30.85%.

(b) Proportion$(T = 1)$ = Proportion$(T = 4)$ = 0.05;   Proportion$(T > 2)$ = 0.65.

(c) Proportion$(L > 2)$ = 2/3.

6.73

(a) Proportion$(X > 3)$ = Prop$(Z > (3-2)/2)$ = normalcdf$(3,E99,2,2)$ = 0.3085.

(b) 0.5.

(c) Proportion$(X > 3)$ = 1/8.

(d) 17/25.

6.75

(a) More than 6.

(b) The proportion ot the observations to the right of 6 is much larger than the proportion of observations to the left of 6. Since the median divides the area under the curve in two equal parts, the median must be larger than 6.

(c) 0.5.

6.77

(a) 1/6.

(b) 49%.

(c) p-value = 0.5 (1/6) = 1/12 = 0.083.

(d) Since the p-value is < 0.10, we reject H_0.

6.79

(a)

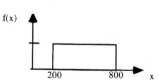

(b) Proportion$(X < 300$ under H_1 $)$ = 0.01.

(c) i.   one-sided to the right.

(d) α = (800 – 700)/600 = 1/6 or 16.7%.

(e) β = 0.5.

(f) p-value = (800 – 400) /600 = 2/3 = 0.67.

(g) No, p-value $> \alpha$.

Interactive Statistics 2nd Edition: Chapter 7 Solutions

7.1
(a) Personal probability.
(b) Relative-frequency approach.

7.3
The long-run relative frequency approach may not be appropriate because not all events are repeatable.

7.5
Answers will vary.

7.7
(a) Mrs. Smith is right.
(b) Answers will vary.

7.9
(a) With the calculator assign 1 as the winning number and 2, 3, 4, and 5 as the losing numbers. With the random number table we can assign 01 – 20 as the winning numbers and the other remaining two digit numbers to be the losing numbers.
(b) With the calculator we get the numbers:
    4, 4, 2, 4, 3, 1, 4, 4, 1, 1,   1, 2, 3, 5, 2, 5, 5, 2, 1,5 ,
    5, 2, 1, 1, 2, 2, 5, 4, 4, 4,   5, 4, 5, 3, 2, 1, 4, 4, 1, 3,
    2, 1, 1, 4, 1, 5, 1, 3, 1, 5.
    The first five numbers obtained with the random number table are: 91, 57, 64, 25, 95; and all represent losses.
(c) With the calculator we have: 14/50 = 0.28.

7.11
(a) S = { EEE, EEG, EGE, GEE, EGG, GEG, GG}.
(b) A = {GEG, EGE}.

7.13
(a) $S = \{F, P\}$.
(b) $S = \{$FFFF, FFFP, FFPF, FFPP, FPFF, FPFP, FPPF, FPPP, PFFF, PFFP, PFPF, PFPP,PPFF, PPFP, PPPF, PPPP$\}$.
(c) $S = \{0, 1, 2, 3, 4, 5, 6, 7, 8, 9, 10, 11, 12, 13, 14, 15, 16, 17, 18, 19, 20\}$.

7.15
(a) No, since you can have a king of hearts.
(b) Yes, since you cannot have a card that is both a heart and a spade.
(c) No, since you can have a king of spades.

7.17
(a) Error: 0.19 + 0.38 + 0.29 + 0.15 = 1.01 > 1.
(b) Error: 0.49 + 0.52 = 0.92 < 1.
(c) Error: -0.25 < 0.
(d) Error: Probability is 0 since the events "heart" and "black" are disjoint.

7.19

(a) 0.70.

(b)

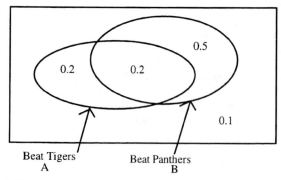

(c) 0.90.

(d) 0.20/0.40 = 0.50.

7.21

(a) 0.40.

(b) 0.30.

(c) 0.25.

7.23

(a) P(D) = 110/1160.

(b) P(D | O) = 20/470.

(c) P(O and D) = 20/1160; P(O) = 470/1160; P(D) = 110/1160.
The two events O and D are NOT independent because P(O and D) ≠ P(O) P(D).

7.25

(a) P(Hypertension) = P(H) = 87/180.

(b) P(Hypertension | Heavy smoker) = 30/49.

(c) No, since P(Hypertension and Heavy smoker) ≠ P(Hypertension) P(Heavy smoker)
That is, 30/180 ≠ (87/180) (49/180).

7.27

(a)

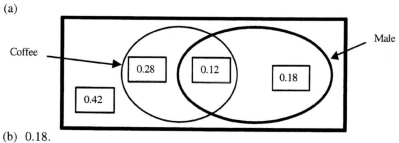

(b) 0.18.

7.29

$1 - (0.18)^2 = 0.9676.$

7.31

(0.6)(0.6)(0.6) = 0.216.

7.33

(a) If A and B are independent events then P(A and B) = (0.4) (0.2) = 0.08.

(b) If A and B are mutually disjoint events then P(A or B) = P(A) + P(B) = 0.6.

7.35

(a) False.

(b) True.

7.37

(a) No. Here is a counterexample. Suppose the sample space is S={1, 2, 3,4, 5}, and let A= {1} and B= {2} be mutually exclusive. Their complements A^c = {2, 3, 4, 5 and B^c = {1, 3, 4, 5} are not mutually exclusive.

(b) Yes. If A and B are independent then P(A and B) = P(A) P(B) then A^c and B^c are independent. Since P(A^c and B^c) = P (A^c) P(B^c).

7.39

(a) 1/3.

(b) 0.7143.

7.41

(a) P(I successful) = 0.7, P(II successful) = 0.3

P(introduce new| I successful) = 0.8

P(introduce new| II successful) = 0.4

P(new) = P(new and I) + P(new and II) = (0.7)(0.8) + (0.3)(0.4) = 0.56+ 0.12 = 0.68.

(b) P(I successful | introduce new) = P(I and new)/ P(new) = 0.8235.

7.43

(a) P(I) = 0.7, P(II) = 0.2, P(need repair | I) = 0.03, P(need repair | II) = 0.04,

P(need repair | I) = 0.05,.

P(repair) = (0.7)(0.03) + (0.2)(0.04) + (0.1)(0.05) = 0.034.

(b) Since they are independent (0.034)(0.034) = 0.001156

(c) P(III | repair) = P(III and repair)/ P(repair) = (0.1)(0.05)/0.034 = 0.147.

7.45

(a) P(1) = 0.7, P(D|1) = 0.02, P(D|2) = 0.05

P(D) = (0.7)(0.02) + (0.3)(0.05) = 0.029.

 (b) P(1|D) = P(1 and D) / P(D) = (0.7)(0.02)/0.029 = 0.482758.

7.47

(a) 0.2.

(b) 0

(c) 0.4.

(d) 0.1/0.4 = 0.25.

7.49

(a) S = {T, WT, WWT}.

(b) X = 1, 2, or 3.

(c) S = {T, WT, WWT, WWWT, ...} so X = 1, 2, 3, 4, ...

7.51

(a) $P(X = 4) = 0.05$.

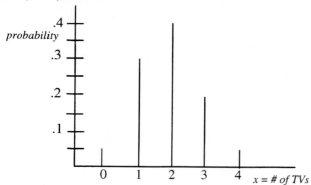

(b) 0.95.

(c) $E(X) = 1.9$.

7.53

(a) $P(X = 0) = P(X = 1) = P(X = 2) = P(X = 3) = 1/4$

(b) $E(X) = 1.5$, σ = standard deviation = 1.118.

7.55

(a) The expected value for Y is one.

(b) The variance of Y is zero.

7.57

(a) $P(X = 4) = 1/6$.

(b) $P(\text{at least six cars}) = 0.833$.

(c) $E(X) = 6.833$.

(d) $E(Y) = 2(6.833) - 1 = 12.666$.

7.59

(a)

How many subsets contain ...	Answer	The subsets are:	Combination:
... exactly 0 values?	1	empty set	$\binom{5}{0} = 1$
... exactly 1 value?	5	{1}, {2}, {3}, {4}, {5}	$\binom{5}{1} = 5$
... exactly 2 values?	10	{1,2}, {1,3}, {1,4}, {1,5} {2,3}, {2,4}, {2,5}, {3,4} {3,5}, {4,5}	$\binom{5}{2} = 10$
... exactly 3 values?	10	{1,2,3}, {1,2,4}, {1,2,5}, {1,3,4}, {1,3,5}, {1,4,5}, {2,3,4}, {2,3,5}, {2,4,5}, {3,4,5}	$\binom{5}{3} = 10$
... exactly 4 values?	5	{1,2,3,4}, {1,2,3,5}, {1,2,4,5}, {1,3,4,5}, {2,3,4,5}	$\binom{5}{4} = 5$
... exactly 5 values?	1	{1,2,3,4,5}	$\binom{5}{5} = 1$

(b) $32 = 1 + 5 + 10 + 10 + 5 + 1$.

7.61
(a) Yes. Each day you win or you don't win, and the probability of winning remains the same all the time, and the outcomes are independent to each other.
(b) No. The responses are not independent.

7.63
(a) H_0: 1% of the tax returns were audited, $p = 0.01$
   H_1: More than 1% of the tax returns were audited, $p > 0.01$
(b) We have $X \sim \text{Bin}(18, 0.01)$, p-value $= P(X \geq 1) = 1 - P(X = 0) = 1 - 0.8345 = 0.1655$
(c) The p-value is the probability of observing exactly one or more tax audited returns among 17, if indeed 1% of tax returns are audited. Or, If we were to take many, many samples of size 18 from this population, we can expect to see 16.55% of them to have 1 or more audited tax returns, if indeed 1% of tax returns are audited.
(d) You would accept the null hypothesis.

7.65
(a)

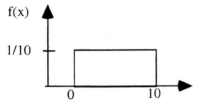

(b) $E(X) = 5$.
(c) 3/10.
(d) (i) to the right
   (ii) p-value = 1/10
   (iii) We reject H_0 since the p-value = α.

7.67
(a) $P(T < 18) = 0.1151$.
(b) $(200)(0.1151) =$ approximately 23.

7.69
(a) $P(1050 < X < 1250)$ is the probability that a randomly selected applicant has a SAT score between 1050 and 1250. $P(1050 < X < 1250) = 0.7745$.
(b) i. 97.5%.
   ii. About 1013.

7.71
{DDD, DDN, DND, NDD, NND, NDN, DNN, NNN}.

7.73
(a) $0.12 + 0.10 - 0.17 = 0.05$.
(b) 0.078.

7.75
(a) P(you win) = 0.10.
(b) P(you win) = 0.41.

7.77
(a) 0.60.
(b) 0.35.
(c) 0.18.
(d 0.30.
(e) 0.60.
(f) No. $0.18 \neq (0.60)(0.35)$.

7.79
(a) 0.30.
(b) 5/11.
(c) No. The events can occur at the same time.
(d) No. P(Nonsmoker and 0 visit) = 0.25 $\neq$ P(Nonsmoker) P(0 visit) = (11/20)(0.3).

7.81
(a) No, the events are not independent because
$$P(\text{F attend and S attend}) = \frac{18}{80} = 0.225 \neq P(\text{F attend})P(\text{S attend}) = \left(\frac{25}{80}\right)\left(\frac{40}{80}\right) = 0.15625.$$
(b) No, the events are not mutually exclusive because there are 18 families in the intersection.

7.83
The answer is (a).

7.85
(a) Not independent.
(b) Not mutually exclusive.
(c) 0.40.

7.87
P(at least one likes statistics) = 1 - P(neither one likes statistics)
= 1 - P(both don't like statistics) = $1 - (.40)^2$ = 1-0.16 = 0.84

7.89
Let M1 = first machine, M2 = second machine, M3 = third machine, and D = defective item.
$P(M1) = 0.50$, $P(M2) = 0.30$, $P(M3) = 0.20$, $P(D \mid M1) = 0.04$, $P(D \mid M2) = 0.06$, and $P(D \mid M3) = 0.02$.
So $P(D) = (0.50)(0.04) + (0.30)(0.06) + (0.20)(0.02) = 0.042$.

7.91
(a) False.
(b) True.
(c) False.
(d) False.

7.93
(a) $P(X = 4) = 0.05$.
(b) 0.85.
(c) 0.25.
(d) $E(X) = 1.65$.

7.95
(a) 0.80.
(b) $E(X) = 2.4$.

7.97

(a)

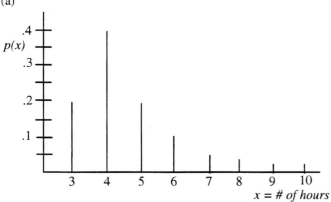

$x = $ # of hours

(b)   Mode = 4.
(c)   E(X) = 4.58.
(d)   Less than.
(e)   0.70.
(f)   0.75.
(g)   Response bias.

7.99

(a)   0.60.
(b)   E(X) = 3.19.

7.101

First note that P(at least 1 has blood type O) = 1 - P(none have blood type O). Let X be the random variable that takes on the number of people that have blood type O in a random sample of n=4, $X \sim \text{Bin}\,(4, 0.4)$.
$$P(X \geq 1) = 1 - P(X = 0) = 1 - 0.1296 = 0.8704$$

7.103

(a)   0.37580.
(b)   0.00569.

7.105

(a)   P(X > 592) = 0.69.
(b)   0.478.
(c)   They are not independent events.

8.1

(a) The sampling distribution of a statistics is the distribution of the values of the statistic in all possible samples of the same size from the same population. We often generate the empirical sampling distribution using some form of simulation.

8.3

The answer is (b).

8.5

(a) Histogram C because X takes on only two values, 0 and 1, with probabilities of 0.80 and 0.20.
(b) Histogram B because we know with a large sample size the distribution of the sample proportion will be approximately normal, centered at the true $p = 0.20$.

8.7

(a) Yes.
(b) Yes.
(c) No, since $np < 5$.

8.9

Note that $\hat{p} \approx N(0.50, 0.025)$ so $P(\hat{p} \geq 0.45) = P(Z \geq 2) = 0.0228$;
or normalcdf(-E99, 0.45, 0.5, 0.025) = 0.02275.

8.11

Answer is (c).

8.13

(a) Approximately normal with mean of 0.85 and standard deviation of 0.0357.
(b) The sample proportion is 90/100 or 0.90.
(c) p-value = $P(\hat{p} \geq 0.85) = P(Z \geq 1.40) = 0.0808$; or normalcdf(0.85, E99, 0.85, 0.0357) = 0.0808.
(d) Since the p-value is more than 0.05, the new method is not significantly more accurate at 5% significance level.

8.15

(a) The true standard deviations are given by:
   $n = 50$, $p = 0.1$ standard deviation = 0.0424; $n = 50$, $p = 0.3$ standard deviation = 0.0648;
   $n = 50$, $p = 0.7$ standard deviation = 0.0648;
   $n = 50$, $p = 0.5$ standard deviation = 0.0707; $n = 100$, $p = 0.5$ standard deviation = 0.050.
(b) Answers will vary.
(c) Answers will vary.

8.17

False. The central limit theorem states that for large sample size, the distribution of $\overline{X}$ is approximately normal, it will not be exactly normal. The approximation is better for larger the sample size.

8.19

(a) Since X is $N(1250, 150)$, the probability is

$$P(1200 < X < 1400) = P\left(\frac{1200-1250}{150} < Z < \frac{1400-1250}{150}\right) = P(-.33 < Z < 1) = 0.4719$$

or with normalcdf(1200, 1400, 1250, 150) = 0.4719.

(b) Since $\overline{X}$ is $N(1250, 150/6)$, the probability is

$$P(1200 < \overline{X} < 1400) = P\left(\frac{1200-1250}{150/\sqrt{36}} < Z < \frac{1400-1250}{150/\sqrt{36}}\right) = P(-2 < Z < 6) = 0.9772$$

or with normalcdf(1200, 1400, 1250, 25) = 0.9772.

(c) The probability is part (b) is higher since the distribution of $\overline{X}$ is more concentrated around 1,250 (i.e. has a smaller standard deviation) than that of X.

8.21

(a) $P(X > 240) = P\left(Z > \frac{240-170}{30}\right) = P(Z > 2.33) = 0.0098$ or with normalcdf((240, E99,170, 30) = 0.0098.

(b) $\overline{X}$ is $N(170, 30/4)$ so the standard deviation is 7.5.

(c) $P(\overline{X} > 190) = P\left(Z > \frac{190-170}{7.4}\right) = P(Z > 2.67) = 0.0038$ or with normalcdf(190, E99, 170, 30/4) = 0.0038.

8.23

Fail to operate implies that the total weight exceeds 5000 or equivalently that the average or mean weight exceeds

555.6. The probability is $P(\overline{X} > 555.6) = P\left(Z > \frac{555.6-540}{45/\sqrt{9}}\right) = P(Z > 1.04) = 0.1498$

or using normalcdf(555.6, E99, 540, 15) = 0.1498.

8.25

(a) Histogram C.

(b) Histogram F.

8.27

(a) 0.61

(b) $E(X) = 1.2$

(c) $\overline{X} \sim N(1.2, 0.9/100)$.

(d) $P(\overline{X} < 1) =$ normalcdf(-E99,1,1.2,.9/100) = 0.

8.29

Note that the total number of tickets desired being less than or equal to 220 is equivalent to having the sample mean number of tickets desired for 100 students being less than or equal 2.2. The sample mean has a $N(2.1, 0.2)$ distribution and the probability is given by: $P(\overline{X} < 2.2) = P(Z < 0.5) = 0.6915$ or use normalcdf(-E99, 2.1, 2.2, 0.2).

8.31

(a) False.

(b) True.

(c) False.

(d) False.

8.33

(a) The answer is D.

(b) No. If the variability remains the same, the sampling distribution does not depend on the population size.

8.35

The population proportion is $p = 0.60$.

So the sample proportion $\hat{p} \approx N(0.6, \sqrt{.24/100})$, that is, $\hat{p} \approx N(0.6, 0.049)$.

$P(\hat{p} < 0.5) = P(Z < (0.5 - 0.6)/0.049) = 0.0206$, or using normalcdf(-E99, 0.5, 0.6, 0.049) = 0.0206. Thus it is very unlikely to observe a $\hat{p} < 0.50$ for a random sample of size 100 if the true proportion is 0.60.

8.37

(a) The three remaining possible samples are: (9,1), (9,5), (9,9). The sample means for all nine possible samples are: 1, 3, 5, 3, 5, 7, 6, 7, 9.

(b) 3/9.

(c) 5.

8.39

(a) The mean is 4 and the variance is 8.

(b) Using TI with Seed = 83   Using random number table with row 10, column 1

Number 1	Number 2	Mean	Number 1	Number 2	Mean
8	0	4	8	4	6
6	8	7	6	8	5
8	0	4	4	2	3
6	6	6	8	8	8
4	8	6	0	6	3
8	6	7	0	8	4
6	4	5	8	6	7
2	4	3	6	2	4
2	8	5	0	0	0
8	8	8	0	8	4
8	6	7	8	8	8
0	4	2	6	4	5
8	6	7	4	8	6
8	4	6	8	6	7
4	2	3	4	2	3
4	4	4	2	8	5
2	0	1	8	6	7
2	0	1	8	8	8
8	2	5	8	2	5
8	8	8	2	6	4
6	2	4	0	6	3
8	8	8	6	8	7
8	0	4	6	0	3
4	4	4	8	0	4
2	2	2	0	6	3

(c) Here is the frequency plots of the sample means of part (b):

```
TI:                          Random Number Table:
X                            X
X                            X X
X       X                    X X X   X
X X X X X                    X X X   X X
X X X X X X X X              X X X X X
X X X X X X X X        X     X X X X X X
--------------------------   --------------------------
0 1 2 3 4 5 6 7 8            0 1 2 3 4 5 6 7 8
   Sample means                 Sample means
```

(d) Using TI with Seed = 124

Number 1	Number 2	Number 3	Number 4	Number 5	Mean
4	0	8	0	0	2.4
4	8	0	6	8	5.2
2	6	4	6	2	4.0
8	6	4	8	8	6.8
8	2	4	8	4	5.2
0	8	6	4	6	4.8
6	6	4	4	4	4.8
2	0	0	0	0	0.4
2	8	4	4	4	4.4
0	4	4	0	2	2.0
0	0	8	2	0	2.0
6	0	2	4	6	3.6
8	2	2	6	2	4.0
2	0	2	6	8	3.6
4	2	6	4	2	3.6
8	2	6	0	0	3.2
8	6	0	6	2	4.4
4	2	6	2	8	4.4
6	0	4	8	2	4.0
4	4	2	8	8	5.2
0	2	0	6	2	2.0
0	4	0	2	2	1.6
0	6	0	2	6	2.8
0	4	0	8	8	4.0
8	4	6	6	2	5.2

Using the random number table with row 20, column 1:

Number 1	Number 2	Number 3	Number 4	Number 5	Mean
0	0	6	6	2	2.8
8	8	0	8	4	5.6
2	6	8	0	6	4.4
6	6	8	8	6	6.8
0	2	8	4	6	4.0
0	4	8	8	6	5.2
0	2	6	2	8	3.6
0	4	4	8	6	4.4
6	2	4	8	8	5.6
2	8	4	4	6	4.8
0	2	0	6	8	3.2
0	2	2	0	4	1.6
4	2	2	8	0	3.2
4	2	6	6	6	4.8
6	0	4	2	8	4.0
2	4	6	4	8	4.8
4	2	2	2	4	2.8
2	4	0	4	0	2.0
0	2	0	6	6	2.8
4	6	8	2	6	5.2
8	4	8	0	6	5.2
6	4	6	2	2	4.0
4	2	2	6	0	2.8
4	4	2	2	6	3.6
2	6	0	2	4	2.8

(e) Here are the frequency plots of the sample means of part (d):
Note: The X's above the 2 represent values in the interval (2.0, 3.0].
TI: Random Number Table:

```
      X
      X
      X
      X X                      X
  X  X X X                     X
  X  X X X                 X X X X
  X X X X X                X X X X
  X X X X X X X            X X X X
                           X X X X X
                           X X X X X X
  -------------------      -------------------
  0 1 2 3 4 5 6 7 8        0 1 2 3 4 5 6 7 8
    Sample Means             Sample Means
```

(f) The second plots have less variation about the mean. The data are more clustered around the true mean of 4.

8.41

(a) $1 - (0.39 + 0.24 + 0.14 + 0.09 + 0.05 + 0.03) = 0.06$.

(b) Histogram C.

8.43

(a) $P(X > 180) = P\left(Z > \dfrac{180 - 173}{30}\right) = P(Z > 0.2333) \approx 0.4080$

or use normalcdf(180,E99,173,30) = 0.4078.

(b) $P(\overline{X} > 180) = P\left(Z > \dfrac{180 - 173}{30 \big/ \sqrt{36}}\right) = P(Z > 1.4) = 0.0808$.

or use normalcdf(180,E99,173,5) = 0.0808.

8.45

(a) (iii) Can't Tell.

(b) $\overline{X} \sim N(-3, 4/20)$ so the standard deviation is 0.20.

(c) $P(\overline{X} < -2.8) = P\left(Z < \dfrac{-2.8 - 3}{0.20}\right) = P(Z < 1) = 0.8413$ or use normalcdf(-E99, -2.8, -3, 0.2) = 0.841.

8.47

No, its variance is smaller.

8.49

Pablo is correct - as the size of each sample used increases, the sampling distribution has less variability. Eduardo is also correct, the value of the parameter p does influence the variability for the sampling distribution of a sample proportion.

9.1

(a) H_0: The proportion of democrats in Wayne County is equal to 0.50.

H_1: The proportion of democrats in Wayne County is more than 0.50.

(b) H_0: The proportion of male births in Wayne County is equal to the proportion of male births in Oakland County.

H_1: The proportion of male births in Wayne County is not equal to the proportion of male births in Oakland County.

9.3

(d) It is equal to α.

9.5

(b) p-value = 0.005.

9.7

(a) The population of pregnant women who work with a computer 1-20 hours per week.

(b) The test statistic formula is $Z = \dfrac{\hat{p} - p_0}{\sqrt{\frac{p_0(1-p_0)}{n}}}$. The observed test statistic value is $z = 1.4772$. We cannot reject H_0 because the p-value = 0.0698 > 0.01.

(c) No, the results are not statistical significant at the level 0.01.

9.9

(a) $\hat{p} = 0.7289$.

(b) The test can be performed using the 1-PropZTest with the calculator. The test statistic is $z = 1.323$ and the corresponding p-value is $P(Z \geq 1.323) = 0.0930$. We cannot reject the null hypothesis.

9.11

H_0: $p = 0.2$ versus H_1: $p > 0.2$; $\hat{p} = 0.72$. The p-value = $P(Z \geq 9.19) = 0$. Reject H_0.

9.13

H_0: The proportion of all reports with discrepancies = 0.20

H_1: The proportion of all reports with discrepancies < 0.20

The test statistic value = -1.5811 and p-value = 0.0568. We cannot reject H_0 at the level 0.05. The results are not statistically significant at 0.05.

9.15

(a) It may be a sample of typical working mothers with young children.

(b) H_0: $p = 0.5$ versus H_1: $p \neq 0.50$.

(c) No. We do not know how many of the 1200 had children under six.

9.17

Reject H_0 if $Z \geq 2.576$ or $Z \leq -2.576$.

9.19

(a) $\hat{p} = 17/40 = 0.425$

(b) The 99% confidence interval for the population proportion is (0.22367, 0.62633). It can be found from the 1-PropZInt with the calculator or using the confidence interval formula with $z^* = 2.576$.

(c) The margin of error = $E = 0.20133$.

9.21

(a) The 95% confidence interval for the population proportion is (0.66784, 0.73216)

(b) The population proportion p represents the true proportion of operating vehicles that are equipped with air bags.

9.23

(a) 633

(b) A statistic.

(c) $\hat{p} = 0.63$.

(d) Standard error of the sample proportion is: $\sqrt{\dfrac{(0.63)(0.31)}{1004}} = 0.0139$

(e) The 95% confidence interval is (0.60062, 0.66033)

(f) The margin of Error = E= 0.02986 ≈ 0.03. Yes, it is close to the stated value.

9.25

The 92% confidence interval is (0.0151, 0.1099).

9.27

A confidence interval is constructed for estimating a population parameter.

9.29

(a) $\hat{p}$ is approximately $N(p, \sqrt{\dfrac{p(1-p)}{n}})$ where $n = 4000$.

(b) No.

(c) Yes, the sample size is in the denominator of the variance.

(d) The confidence interval of (0.51, 0.55) was made with a method, which if repeated, would result in many 95% confidence intervals. We would expect 95% of these possible intervals to contain the population proportion p.

9.31

(a) The interval provides plausible values for the true population proportion of abused children with above average anxiety level, based on this sample of 60 such children.

(b) If we repeat this procedure many times, we would expect the true proportion p to lie inside about 95% of the confidence intervals produced in this manner.

9.33

A sample of at least 1504 students is needed.

9.35

(a) (i) – (iv) are false. (v) True.

(b) $n \geq (1.96/0.08)^2 = 601$

9.37

(a) A statistic.

(b) (0.79199, 0.92801)

(c) E = 0.06801.

(d) $n \geq (1.96/0.08)^2 = 601$.

(e) H_0: $p = 0.87$ versus H_1: $p < 0.87$.

(f) The test statistic value is $z = -0.2974$. The p-value = 0.3831. You cannot reject H_0 at the 0.05 level.

9.39

H_0: $p = 2/3$ versus H_1: $p \neq 2/3$.

9.41

(a) H_0: $p = 0.72$ versus H_1: $p > 0.72$. One-sided to the right.

(b) H_0: $p = 0.90$ versus H_1: $p < 0.90$. One-sided to the left.

(c) H_0: $p = 0.60$ versus H_1: $p \neq 0.60$. Two-sided.

9.43

(a) No, since we do not know the number of pregnant women.

(b) (0.1344, 0.1656).

(c) Approximately 100.

9.45

(a) The results are statistically significant at the 5% level.

(b) Sometimes.

9.47

(a)   (0.491, 0.549).

(b) The margin of error is 0.029, that is approximately 2.9%.

9.49

(a) H_0: $p = 0.50$ versus H_1: $p > 0.50$.

(b) $z = 1.7244$; p-value $= 0.0423$.

(c) Yes, the p-value is smaller than 0.05.

9.51

(a) H_0: $p = 0.60$ versus H_1: $p > 0.60$.

(b) The test statistic value $= 1.4434$. The p-value $= 0.0745$. The results are not statistically significant. We cannot reject H_0 at the 0.05 level.

(c) No. The sample size should be large.

9.53

(a) (0.44126, 0.51874).

(b) The margin of error $= E = 0.03874$

(c) The margin of error will increase to 0.06066.

9.55

Answers will vary.

Interactive Statistics 2nd Edition: Chapter 10 Odd Solutions

10.1
(a) H_0: $\mu = 33$ °F versus H_1: $\mu < 33$ °F.
(b) H_0: $\mu = 26$ years versus H_1: $\mu > 26$ years.
(c) H_0: $\mu = 200$ versus H_1: $\mu \neq 200$.

10.3
(a) H_0: $\mu = 5000$ versus H_1: $\mu < 5000$.
(b) $z = -2.5$.
(c) p-value $= 0.0062$.
(d) Yes, since the p-value < 0.01.

10.5
H_0: $\mu = 60$ versus H_1: $\mu < 60$, the test statistic $= z = -2.5$, p-value $= 0.0062$. We reject H_0.

10.7
(a)

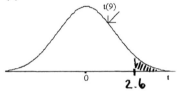

Using the calculator the p-value $=$ tcdf(2.6, E99, 9) $= 0.014369$. Using the t-distribution table the bounds for the p-value are given by: $0.01 < p$-value < 0.025.
(b)

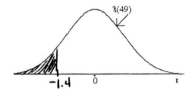

Using the calculator the p-value $=$ tcdf(-E99, -1.4, 49) $= 0.0839$. Using the t-distribution table the bounds for the p-value are given by: $0.05 < p$-value < 0.10.
(c)

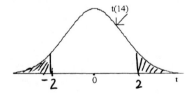

Using the calculator the p-value $=$ 2tcdf(-E99, -2, 14) $= 2(0.03264) = 0.06529$. Using the t-distribution table the bounds for the p-value are given by: $0.05 < p$-value < 0.10.

10.9

(a) H_0: $\mu = 16$ versus H_1: $\mu \neq 16$.

(b) Assume the distribution of the amount of syrup is normal.

(c) The company would turn off the equipment to investigate and lose money.

(d) The company either loses money or customers would be dissatisfied.

(e) Using the calculator with option 2:TTest we have: $t = -1.28$, p-value $= 0.2167$. Using the t-distribution table we have that the p-value is $0.2 < $ p-value $ < 0.25$ So we would not reject the null hypothesis.

(f) Our advice to the company is to let the machine continue to run.

10.11

(a) Using the calculator with option 2:TTest we have $t = 1.46$; p-value $= 0.0858$. Using the t-distribution table we have that $0.05 < p$-value < 0.10.

(b) Since the p-value is less than 0.10, we reject the null hypothesis.

(c) Yes, a Type I error. The probability of having committed an error is either 0 or 1.

10.13

(a) H_0: $\mu = 12$ versus H_1: $\mu > 12$.

(b) Using the calculator with option 2:TTest we have: $t = 1.3765$, p-value $= 0.1136$. Using the t-distribution table we have that $0.10 < p$-value < 0.20.

(c) Assume the data are a random sample and concentration has a normal distribution.

(d) If the mean concentration were 12 mg/kg, we would see a test statistic of 1.376 or even larger about 11.4% of the time in repeated sampling.

(e) Do not reject the null hypothesis since the p-value is $> \alpha$.

10.15

(a) Using the calculator with option 2:TTest we have: $t = 0.475$, p-value is $= 0.323$. Using the t-distribution table we have that $0.3 < p$-value < 0.4.

(b) Since the p-value is $> \alpha$, we cannot reject the null hypothesis.

10.17

H_0: $\mu = 50$ versus H_1: $\mu < 50$.
Using the calculator with option 2:TTest we have: $t = -2.4$, p-value $= 0.0109$. Using the t-distribution table we have that $0.01 < p$-value < 0.05. So we reject H_0 at $\alpha = 0.05$.

10.19

(a) $\bar{x} = 13.5$.

(b) Using the calculator with option 8:TInterval we have: (13.094, 13.906). The value of t^* from the table would be the conservative 2.042 (using 30 degrees of freedom).

(c) If we repeat this method over and over, yielding many 95% confidence intervals for μ, we expect 95% of these intervals to contain the true parameter value.

10.21

(a) Using the calculator with option 8:TInterval we have: (434.99, 675.41). With the t-distribution table we would use $t^* = 2.093$.

(b) The margin of error $= E = 120.21$.

(c) If we repeat this method over and over, yielding many 95% confidence intervals for the true mean, we expect about 95% of these intervals to contain μ.

10.23

The sample mean is 49.2 and the margin of error is $E = 5$.

10.25

Student 4's answer is correct.

10.27

Since σ is not known we use the *t*-distribution. Using the calculator with option 8:TInterval we have: (-3.282, 7.502). With the *t*-distribution table we would use $t^* = 2.262$.

10.29

(a) Using the calculator with option 8:TInterval we have: (75.9,84.1). With the *t*-distribution table we would use $t^* = 1.740$.
(b) $E = 4.1$
(c) No, 82 is inside the confidence interval.
(d Narrower.

10.31

(a) Yes.
(b) Can't tell.
(c) Can't tell.
(d) No.
(e) Yes.

10.33

The correct answer is (b).

10.35

(a) H_0: $\mu = 17.1$ versus H_1: $\mu < 17.1$.
(b) Using the calculator with option 2:TTest we have: $t = -2.3159$, *p*-value $= 0.0229$. Using the *t*-distribution table we have: $0.01 < p\text{-value} < 0.025$.
(c) We reject the null hypothesis. It does appear that the new species has a faster germination time on average.

10.37

(a) $P(X \leq 285) = P(Z \leq (285 - 270)/10) = \text{normalcdf}(-E99, 285, 270, 10) = 0.93319$.
(b) (i) P(at least one among the four pregnancies lasts more than 285 days) =
          $1 - P(\text{all pregnancies last less or equal to 285 days}) = 1 - 0.933191^4 = 0.2416$.

(ii) $P(\overline{X} < 285) = P(Z < (285-270)/(10/2) = \text{normalcdf}(-E99, 285, 270, 5) = 0.99865$.
(c) H_0: $\mu = 270$ versus H_1: $\mu < 270$.
(d) $t = -1.3333$, *p*-value $= 0.101858$.
(e) You cannot reject the null hypothesis at the level 0.05.
(f) Women who work outside the home do not appear to have a different mean pregnancy length than for the general population of women.

10.39

(a) Stratified random sampling.
(b) Using the calculator with a seed value of 331 the selected male students are: 35, 8, 4, 38, 23, 17, 36, 13, 6, 24. Using the random number table (row 18, column 1) the selected male students are:
     1, 15, 40, 33, 36, 29, 4, 31, 27, 30.
(c) (i) $\overline{x} = 7.18$, s = 1.08.
     (ii) Min = 4.2; $Q_1 = 6.6$; Median = 7.1; $Q_3 = 7.9$; Max = 9.2.
(d) H_0: $\mu = 7.5$ versus H_1: $\mu < 7.5$.
     Using the calculator with option 2:TTest we have: $t = -1.3254$; *p*-value $= 0.1004$. Using the *t*-distribution table we have: $0.10 < p\text{-value} < 0.20$.
(e) Since the *p*-value > 0.05, we cannot reject the null hypothesis. The data are not statistically significant at the level 0.05.
(f) $(40/80)(6.82) + (40/80)(7.18) = 7$.

10.41

(a) Min = 230; $Q_1 = 250$; Median = 265; $Q_3 = 290$; Max = 310.

(b) With the calculator you would use the following window:

Xmin =230
Xmax=330
Xscl=20
Ymin=-1
Ymax=8
Yscl=1
Xres=1
Histogram

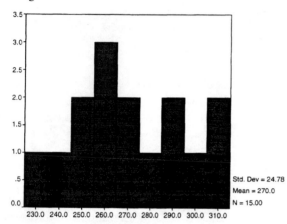

PRICE

(c) H_0 : $\mu = 258$ versus H_1 : $\mu > 258$.

Using the calculator with option 2:TTest we have: $t = 1.875$; p -value = 0.0408. Using the t -distribution table we have: 1.761. We would reject the null hypothesis since $0.0408 < 0.05$. The data are statistically significant.

10.43

(a) Using the calculator with option 8:TInterval we have: (625.35,658.65). With the t -distribution table we would use $t^* = 1.753$.

(b) (i) False.
    (ii) False.

10.45

(a) Using the calculator with option 7:ZInterval we have (11.397, 11.603). With the standard normal table we would use $z^* = 2.576$.

(b) 0.103.

10.47

(a) The apartment in the suburb.

(b) Living area of the apartments in square feet.

(c) 1325 square feet.

(d) Using the calculator with option 8:TInterval we have: (1316.7, 1333.3). With the t -distribution table we would use the formula with a value of $t^* = 2.000$ (using the conservative 60 degrees of freedom).

(e) No. The true mean is or is not inside the interval.

(f) Increase the confidence level or decrease the sample size.

10.49
(a) False.
(b) False.
(c) False.
(d) True.
(e) Accept H_0.

10.51
(a) 7.4.
(b) Accept H_0.
(c) Accept H_0.

Interactive Statistics 2nd Edition: Chapter 11 Odd Solutions

11.1
Paired - Each female twin is directly linked to her twin brother.

11.3
Paired - Each woman is directly related (linked) to her husband.

11.5
(a) The male students can be paired up in the following pairs: Ronald & Lee and Kyle & Pablo; Ronald & Kyle and Lee & Pablo; Ronald & Pablo and Lee & Kyle. The female students can be paired up in the following pairs: Kerri & Emily and Sara & Sonya, Kerri & Sara and Emily & Sonya; Kerri & Sonya and Emily & Sara. One of the 3 possible pairings for males can be combined with any of the 3 possible pairings for females, resulting in nine possible pairings. One example is Ronald & Kyle, Kerri & Sonya, Lee & Pablo, and Emily & Sara.

(b) Ronald & Kyle (20), Kerri & Sonya (18), and then one of the three possible pairs for 19 year olds: Emily & Lee with Pablo & Sara, Emily & Pablo with Lee & Sara, or Emily & Sara with Lee & Pablo.

(c) One possible assignment: Let Ronald = 1, Lee = 2, Kyle = 3, Pablo = 4. Also let Kerri = 1, Emily = 2, Sara = 3, Sonya = 4. One possible method: Step 1: Select 2 random numbers between 1 and 4. These two will form the first male pair. The remaining two form the second male pair. Step 2: Repeat Step 1 for the females. Using the TI: Step 1 (seed = 18): 2, 2 (skip), 1. Male pairs: Ronald & Lee, Kyle & Pablo. Step 2 (seed = 33): 3, 1. Female pairs: Kerri & Sara, Emily & Sonya. Using the random number table: Step 1 (row 8, column 1): 3, 1. Male pairs: Ronald & Kyle, Lee & Pablo. Step 2 (row 22, column 6): 2, 3. Female pairs: Emily & Sara, Kerri & Sonya.

(d) One possible assignment: Let Emily = 1, Lee = 2, Pablo = 3, Sara = 4. One possible method: Step 1: The two 20-year students automatically form pair 1: Ronald & Kyle. The two 18-year old students automatically form pair 2: Kerri & Sonya. Step 2: Select 2 random numbers between 1 and 4. These two will form the first 19-year old pair. The remaining two form the second 19-year old pair. Using the TI: Step 2 (seed = 28): 4, 4 (skip), 2. 19-year old pairs: Lee & Sara, Emily & Pablo. Using the random number table: Step 2 (row 10, column 1): 4, 3. 19-year old pairs: Pablo & Sara, Emily & Lee.

11.7
(a) $H_1 : \mu_D > 0$.

(b) The data are a random sample from a normal population with unknown population standard deviation.

(c) From the table we find the sample mean of the differences is 2 and the sample standard deviation of the differences is 3.464. So the observed test statistic is $t = \dfrac{2}{\dfrac{3.464}{\sqrt{7}}} = 1.5275$. The p-value is

$P(T \geq 1.5275) = 0.0887$, where T has $n-1 = 6$ degrees of freedom. We accept H_0 at the 5% level and conclude that the new program does not seem to improve creative thinking. With the TI we would use the 2: TTest option.

(d) The correct answer is (i).

11.9

(a) Let $\mu_D = \mu_1 - \mu_2$. The hypotheses are: $H_0 : \mu_D = 0$ vs. $H_1 : \mu_D \neq 0$.

(b) The data are a random sample from a normal population with unknown population standard deviation.

(c) There is only one pair (the third pair) for which the highest weight is on scale 2.

(d) Excluding the third pair, we find from the table we find the sample mean of the differences is 2.6 and the sample standard deviation of the differences is: 2.302. The critical value we need is $t* = 2.776$. So the 95% confidence interval is given by: $2.6 \pm 2.7764\left(\dfrac{2.6}{\sqrt{5}}\right)$, or (-0.258, 5.458). With the TI we would use the 8: TInterval option.

(e) The answer is (ii). The p-value would be greater than 0.05, since 0 is included in the interval, so we accept the null hypothesis.

11.11

(a) The observed sample mean is 1.45 standard errors above the hypothesized mean of 0.

(b) The p-value would be $0.190/2 = 0.095$.

(c) No, we would not reject the null hypothesis, because the p-value is greater than α.

(d) No, because the alternative hypothesis is one-sided and a confidence interval can only be used for a two-sided test.

11.13

(a) $H_0 : \mu_D = 0$ versus $H_1 : \mu_D > 0$ where D = second score – first score.

(b) This is a paired t-test so we need to assume that the differences are a random sample from a normal population.

(c) The standard error of the mean difference = $\dfrac{s_D}{\sqrt{n}} = \dfrac{41.4}{\sqrt{4}} = 20.7$. We estimate the average distance of the possible sample mean differences from μ_D to be about 20.7 points.

(d) We have $\overline{d} = 100/4 = 25$. So the observed test statistic is $t = 25/20.7 = 1.208$. Using Table IV: $0.15 < p\text{-value} < 0.20$. Using the TI: $p\text{-value} = 0.157$. Decision: accept H_0. We conclude that there is not sufficient evidence to say students improve their score the second time they take the SAT.

(e) ii ... about 5% of the time.

11.15

(a) The differences are: 7, 6, -1, 5, 6, 1

(b) $\overline{d} = 4$

(c) $H_0 : \mu_D = 0$ versus $H_1 : \mu_D > 0$ where D = Before – After.

(d) The population of differences is assumed to be normal.

(e) $t = \dfrac{\overline{d} - 0}{\dfrac{s_D}{\sqrt{n}}} = \dfrac{4 - 0}{\dfrac{3.225}{\sqrt{6}}} = 3.038$

(f) Using Table IV: $0.01 < p\text{-value} < 0.02$. Using the TI: $p\text{-value} = 0.0144$.

(g) Yes

(h) Type I Error

11.17

(a) Common (or equal) population standard deviations.

(b) Yes, the sample standard deviations are similar.

(c) We have $s_p = \sqrt{\dfrac{(16-1)(90)^2 + (16-1)(100)^2}{16+16-2}} = 95.1315$, so the 90% confidence interval is given by (600 – 550) $\pm$ (1.697)(95.1315)$\sqrt{\frac{1}{16} + \frac{1}{16}}$ => (-7.08, 107.08). Using the 0:2-SampTInt option on the TI would yield: (-7.086, 107.09).

11.19

(a) The sample means are 2.173 and 2.523. The sample standard deviations are 0.375 and 0.365. The pooled

sample standard deviation is $s_P = \sqrt{\dfrac{(14)(0.375)^2 + (12)(0.365)^2}{26}} \approx 0.370$ and the 95% confidence

interval for $\mu_1 - \mu_2$ is given by $(2.173 - 2.523) \pm (2.056)(0.370)\sqrt{\dfrac{1}{15} + \dfrac{1}{13}}$, or (-0.6385, -0.0615).

With a TI, you would use the 0:2-SampTInt option.

(b) The interval provides a range of plausible values for $\mu_1 - \mu_2$ at a 95% confidence level. If we repeat this procedure over and over, yielding many 95% confidence intervals for $\mu_1 - \mu_2$, we would expect that approximately 95% of these intervals would contain the true parameter value $\mu_1 - \mu_2$.

(c) You could test the hypotheses $H_0 : \mu_1 = \mu_2$ versus $H_1 : \mu_1 \neq \mu_2$. Since the confidence interval does not include the value of 0, we would reject the null hypothesis.

11.21

(a) The point estimate is the difference in the observed sample means, 1093 - 790 = 303.

(b) We do not know. It will depend on the sample sizes used and the standard deviations of the samples. We would also need check to see if the assumption of common population variance and approximate normal distributions are reasonable.

11.23

(a) The hypotheses are: $H_0 : \mu_1 = \mu_2$ versus $H_1 : \mu_1 \neq \mu_2$.

(b) Strain 1: sample mean is 37.43, sample standard deviation is 3.69.
Strain 2: sample mean is 42.5, sample standard deviation is 3.89.

The pooled sample standard deviation is: $s_P = \sqrt{\dfrac{(6)(3.69)^2 + (7)(3.89)^2}{13}} \approx 3.80$.

The observed test statistic is: $t = \dfrac{(37.43 - 42.5)}{3.80\sqrt{\dfrac{1}{7} + \dfrac{1}{8}}} = 2.579$.

The p-value is $2P(T \leq -2.579) = 0.0229$, where T has 13 degrees of freedom. With a TI, you would use the 0:2-SampTTest option.

(c) Since the p-value is less than 0.05, we reject the null hypothesis and conclude that the average weights for the two strains appear to be different.

(d) (i) Approximately 95% of the intervals produced with this method are expected to contain $\mu_1 - \mu_2$. (ii) We would accept the null hypothesis since the value of -9 is in the confidence interval.

11.25

(a) $H_0 : \mu_1 = \mu_2$ versus $H_1 : \mu_1 > \mu_2$,
where 1 = Students with a C or higher and 2 = Students with D or E.
We have $\bar{x}_1 = 4.6$, $\bar{x}_2 = 2.25$, $s_1 = 3.4$, $s_2 = 1.3$, $s_p = 2.7$, $t = 1.841$, and a p-value = 0.0421. With a TI, you would use the 0:2-SampTTest option. We would reject H_0 and conclude that the Students who earn a C or higher do appear to spend more hours per week outside of class on course work on average as compared to Students who receive a D or E.

(b) Each sample is a random sample from a normal population. The two population standard deviations are assumed to be equal. The two samples are assumed to be independent. The assumption of equal population standard deviation is somewhat suspect, based on boxplots or comparing the sample standard deviations. However, the sample sizes are nearly the same. If we do not assume equal population standard deviations, the test statistic would be t = 2.01 and the p-value would be 0.034, so the same conclusion would be reached. Normality seems reasonable.

11.27

$H_0 : \mu_1 = \mu_2$ versus $H_1 : \mu_1 > \mu_2$, $t = 1.839$, p-value $= 0.073/2 = 0.035$; Reject H_0 and conclude that the Strat method does appear to produce higher scores on average as compared to the Basal method.

11.29

(a) The hypotheses are: $H_0 : p_1 = p_2$ versus $H_1 : p_1 > p_2$, where population 1 is the responses for patients who use carbolic acid and population 2 is the responses for patients who did not use the acid.

(b) The sample proportions are: $\hat{p}_1 = 34/40 = 0.85$ and $\hat{p}_2 = 19/35 = 0.543$. The pooled estimate of p is given by

$$\hat{p} = \frac{34 + 19}{40 + 35} = 0.7067.$$

(c) The observed test statistic is $z = \dfrac{(0.85 - 0.543)}{\sqrt{(0.707)(0.293)\left(\dfrac{1}{40} + \dfrac{1}{35}\right)}} = 2.915$ with a p-value of 0.0018. With the TI

we would use the 6: 2-PropZTest option. Since the p-value is less than 0.05, we reject the null hypothesis and conclude that the presence or absence of carbolic acid does constitute a significant effect. Specifically, patients that use the acid have a better chance of recovery.

11.31

(a) The hypotheses are: $H_0 : p_1 = p_2$ versus $H_1 : p_1 < p_2$, where population 1 is for the year 1992 and population 2 is for the year 1983.

(b) Since the sample proportions are: $\hat{p}_1 = 0.46$ (for 1992, based on a sample of 1250) and $\hat{p}_2 = 0.53$ (for 1983, based on a sample of 1251), we have that the number who avoided excess salt are: $x_1 = (1250)(0.46) = 575$ and $x_2 = (1251)(0.53) \approx 663$. The pooled estimate of p is given by $\hat{p} = \dfrac{575 + 663}{1250 + 1251} = 0.495$. The observed test statistic is $z = \dfrac{(0.46 - 0.53)}{\sqrt{(0.495)(0.545)\left(\dfrac{1}{1250} + \dfrac{1}{1251}\right)}} = -3.50$. The p-value of 0.00023 is less than

0.01, thus we reject the null hypothesis. With the TI we would use the 6: 2-PropZTest option. It appears that there has been a significant decrease in the proportion of Americans who avoid excess salt in their diet.

11.33

$n_1 = n_2 = \frac{1}{2}\left(\frac{1.96}{0.01}\right)^2 \Rightarrow 19{,}208$

11.35

(a) An observational study since no treatment was actively imposed.

(b) Population 1 = Parents of infants who died of SIDS (in Southern California between January 1989 and December 1992); Population 2 = Parents of healthy infants (in the same area and for the same time period).

(c) There is not much detail given. The sample size from each group was 200. It did state that parents of 200 *similar* healthy infants were selected, indicating a possibility of a paired design.

11.37

(a) The sampled differences represent a random sample from a normal population.

(b) The p-value for a one-sided, to the right, alternative hypothesis is $0.06/2 = 0.03$. Since the p-value is less than 0.05, we would reject the null hypothesis. Yes, it appears that the bonus plan did significantly increase sales, on average.

11.39

(a) The 22.3 miles per gallon is the sample mean, because we have a sample of ten old engines, and 22.3 is just the mean of that sample.

(b) In scenario 1, two unrelated samples of units are measured. In scenario 2, related or paired samples are measured.

(c) It would be easier to assess in scenario 2. Related sample are measured, differences in the means between the two samples will better reflect differences between old and new engines and not also differences in driving habits between different drivers.

(d) First rewrite the hypotheses: $H_0 : \mu_1 - \mu_2 = 0$ versus $H_1 : \mu_1 - \mu_2 < 0$. The sample means are 22.3 and 23.1. The sample standard deviations are 2.003 and 3.178. The pooled sample standard deviation is

$$s_P = \sqrt{\frac{(n_1 - 1)s_1^2 + (n_2 - 1)s_2^2}{n_1 + n_2 - 2}} \approx 2.65623 \text{ and the observed test statistic is } t = \frac{(22.3 - 23.1)}{2.66\sqrt{\frac{1}{10} + \frac{1}{10}}} = -0.673.$$

Based on the t-distribution with 9 degrees of freedom, the p-value is 0.2547. With the TI we would use the 4: 2-SampTTest option. Comparing it to $\alpha = 0.05$, we will accept the null hypothesis. At a 5% significance level, the new engines appear to consume as much gas as old engines, on average. We assume we have two independent random samples from normal populations with equal population variance.

(e) The 99% confidence interval for μ_D is given by: $1 \pm (3.2498)\left(\frac{2.044}{\sqrt{10}}\right)$, or (-1.22, 3.22). With the TI we would use the 0: 2-SampTInt option. This interval provides a range of plausible values for μ_D at a 99% confidence level. We expect 99% of such intervals to contain μ_D in repeated sampling.

11.41

(a) Since we have two groups with different sample sizes of 25 and 23, it is not paired, but rather an independent samples design.

(b) This would be a paired design, taking all $n = 25 + 23$ children and recording two measurements for each child.

11.43

(a) The sample proportions are: Treatment group $\hat{p}_1 = 56/2051 = 0.0273$ and Placebo group $\hat{p}_2 = 84/2030 = 0.04137$.

(b) The hypotheses are: $H_0 : p_1 = p_2$ versus $H_1 : p_1 \neq p_2$.

The pooled estimate of p is given by $\hat{p} = \frac{56 + 84}{2051 + 2030} = 0.0343.$

The observed test statistic is $z = \frac{(0.0273 - 0.04137)}{\sqrt{(0.0343)(0.9657)\left(\frac{1}{2051} + \frac{1}{2030}\right)}} = -2.47.$

The two-sided p-value of 0.0135 for assessing if there is a difference is greater than 0.01. Thus we fail to reject the null hypothesis. It appears that there is not a significant difference in the incidence rates of cardiac events for the two groups. With the TI we would use the 6: 2-PropZTest option.

(c) We have (0.0273 - 0.04137)/0.04137 = -0.34, for a 34% reduction.

11.45

(a) 9.74

(b) $H_0 : \mu_D = 0$ (where D = Male – Female) and $H_1 : \mu_D > 0$

(c) $t = 2.216$; p-value = 0.030/2 = 0.015; Reject H_0 and conclude that the male sports do seem to dominate for private schools. With the TI we would use the 2: TTest option.

11.47

(a) H_0: $\mu_1 - \mu_2 = 0$ or $(\mu_1 = \mu_2)$ versus H_1: $\mu_1 - \mu_2 > 0$ or $(\mu_1 > \mu_2)$

(b) $s_p = \sqrt{10,039,260.33} = 3168.48$

(c) $t = 1.748$; p-value $= 0.097/2 = 0.0485$; Reject H_0. With the TI we would use the 4: 2-SampTTest option.

(d) It appears that they do.

(e) No, if we repeated this process many times, we would expect 95% of the confidence intervals generated to contain the true difference in the population means, $\mu_1 - \mu_2$.

Interactive Statistics 2nd Edition: Chapter 12 Odd Solutions

12.1

We test the equality of the means by comparing two estimates of the common population variance σ^2. Both will be unbiased estimators if the means of the populations are equal. If the null hypothesis is not true, then the value of MSB/MSW, i.e., the F-ratio tends to be inflated, because the MSB overestimates σ^2.

12.3

(a) Under H_0, the three populations all have the same normal distribution, with the same common mean and the same common population variance. Under H_1, the three populations all have a normal distribution with at least one mean different from the others, but the variances are the same.

(b) Under H_0, the F-test statistic has an $F(2,30)$ distribution.

12.5

(a) 2.74
(b) 2.99
(c) 5.64
(d) 7.95
(e) 2.33
(f) 1.94

12.7

(a) Since $I - 1 = 3$, there were 4 methods
(b) Since $n - 1 = 60$, there were $n = 64$ observations
(c) Using Table V, the p-value is less than 0.01. Using the TI we have p-value = Fcdf(25.4, E99, 3, 60) = 0.000000000099. The p-value is much less than 0.01, so the result is statistically significant.

12.9

(a) Experiment
(b) Number of Hours without Sleep (or Sleep Deprivation Level).
(c) Number failures (to detect a moving object).
(d) Using the TI: Group 1 = 9, 20, 4, 12, 13; Group 2 = 3, 15, 16, 2, 6; Group 3 = 17, 1, 7, 11, 10; leaving the remaining in Group 4 = 5, 8, 14, 18, 19.
   Using the table we let subject 1 have labels 01, 21, 41, 61, 81 and subject 2 have labels 02, 22, 42, 62, 82, and so on through subject 20 with labels 20, 40, 60, 80, 00. Group 1 = 5, 7, 13, 8, 17; Group 2 = 14, 19, 6, 15, 18; Group 3 = 16, 3, 20, 1, 11; leaving the remaining in Group 4 = 2, 4, 9, 10, 12.
(e) $H_0: \mu_1 = \mu_2 = \mu_3 = \mu_4$ versus H_1: at least one population mean μ_i is different.

12.11

(a) $H_0: \mu_1 = \mu_2 = \mu_3$ versus H_1: at least one population mean μ_i is different. Based on the output, p-value = 0.012 < 0.05, so we reject H_0. There is evidence at the 5% significance level to indicate differences among the mean productions of the three lines.

(b) 50% of the time, Line #1 produced more than 282 or 283 units.

(c) The three independent random samples should be obtained from three normally distributed populations with same variance. By the side-by-side boxplots, the variances seem to be the same. Provided the other assumptions hold, the test appears to be valid.

12.13

(a)

Source	SS	DF	MS	F
Between	104,855	3	34,951.6	9.915
Within	70,500	20	3,525	
Total	175,355			

(b) The estimate of the common population variance is MSW=3525.

(c) $H_0: \mu_1 = \mu_2 = \mu_3 = \mu_4$ versus H_1: at least one population mean μ_i is different.

(d) Using Table V: the p-value is less than 0.01. Using the TI: p-value = Fcdf(9.915, E99, 3, 20) = 0.000325.

(e) Since the p-value is smaller than α, we reject H_0. It appears that not all mean physical fitness scores are the same.

12.15

(a) At the 10% level we accept H_0 since the p-value = 0.117 is larger than $\alpha = 0.10$.

(b) MSW = 3.242.

(c) A large F-ratio means that MSB is overestimating the common population variance. This happens when the population means are not all the same.

12.17

$H_0: \mu_1 = \mu_2 = \mu_3$ versus H_1: at least one population mean μ_i is different, where 1 = Group A, 2 = Group B, and 3 = Group C.

Descriptives

MINUTES

	N	Mean	Std. Deviation
Group A	6	17.000	1.789
Group B	6	14.000	1.414
Group C	6	12.667	1.751
Total	18	14.556	2.431

ANOVA

MINUTES

	Sum of Squares	df	Mean Square	F	Sig.
Between Groups	59.111	2	29.556	10.726	.001
Within Groups	41.333	15	2.756		
Total	100.444	17			

Since the p-value is less than 0.01, we would reject H_0 and conclude that at least one population mean time is different. The three diets do not appear to be the same with respect to the mean time.

12.19

Overall Type I error $= 1 - (1 - \alpha)^c$

c	P(Type I Error)
5	.226
10	.401
25	.723
50	.923
100	.994

12.21

The correct answer is (c).

12.23

(a) H_0: $\mu_1 = \mu_2 = \mu_3$ versus H_1: at least one population mean μ_i is different.

Source	SS	DF	MS	F
Between	223.96	2	111.98	31.11
Within	431.96	120	3.6	
Total	655.92	122		

p-value = $P(F \geq 31.11) \approx 0.000$, where F has $(2,120)$ degrees of freedom. With this very small p-value, we would reject H_0 and conclude that the mean time on market for at least one county is different from the others.

(b) We conclude that there is a significant difference between the mean time on market for County 1 and each of the other two counties (County 2 and County 3).

12.25

		Factor A	
		Level 1	Level 2
Factor B	Level 1	20	30
	Level 2	50	60

12.27

(a)

Fertilizer	Water A	B	C
I	28	31.67	23.33
II	33.67	31	25.67
III	28.67	35.67	29.33
IV	31	31.33	26.67

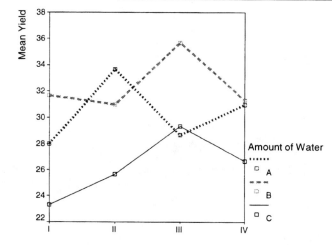

(b) Since the profiles for the three amounts of water are not parallel, there is evidence of interaction. Overall water amount C performs the worst. Amount A works best with Fertilizer II and Amount B works best with Fertilizers I.

12.29

(a) The significance test for interaction is significant with a p-value of 0.006. The significance tests for the two main effects are also significant with p-values less than 0.0005.

(b) Although the profiles are not parallel, they do not cross. The largest difference between the two types of filters appears to occur for medium sized cars, while there is not much of a difference between the filter types for small or large sized cars.

12.31

(a) We accept H_0: $\mu_1 = \mu_2 = \mu_3$ since the p-value of 0.117 is greater than α of 0.10.

(b) MSE $= 3.242$.

12.33

(a) H_0: $\mu_1 = \mu_2 = \mu_3$ versus H_1: at least one population mean μ_i is different.

(b)

Source	SS	DF	MS	F
Between	89.6	2	44.8	5.1
Within	351.5	40	8.79	
Total	441.1	42		

(c) Since the MSW $= 8.79$, the estimate of the standard deviation 2.965.

(d) Using Table V, the p-value is between 0.01 and 0.05.

Using the TI, the p-value is Fcdf(5.1, E99, 2, 40) $= 0.01064$.

12.35

(a) H_0: $\mu_1 = \mu_2 = \mu_3$ versus H_1: at least one population mean μ_i is different

(b)

Level	Sample size	Sample mean	Sample standard deviation
Bottom	3	55.83	2.259
Middle	5	77.60	3.264
Top	5	52.70	2.255

The overall mean is 63.0 and we have $n = 13$ and $I = 3$.

$$SSB = 3(55.83 - 63)^2 + 5(77.6 - 63)^2 + 5(52.7 - 63)^2 = 1750.477$$

$$SSW = 2(2.259)^2 + 4(3.264)^2 + 4(2.255)^2 = 73.161$$

$$MSB = \frac{SSB}{I-1} = \frac{1750.4767}{3-1} = 875.238, \quad MSW = \frac{SSW}{n-I} = \frac{73.161}{13-3} = 7.316$$

$$F = \frac{MSB}{MSW} = \frac{875.238}{7.316} = 119.63. \text{ Since the } p\text{-value is practically 0, we reject } H_0.$$

(c) The 3 populations follow a normal distribution with equal variances. We also need to have a simple random sample from each population. Finally, the 3 random samples are to be mutually independent.

Interactive Statistics 2ⁿᵈ Edition: Chapter 13 Solutions

13.1

(a) $\hat{y} = -30{,}000 + 7{,}000(30) = 180{,}000$.

(b) $\hat{y} = -30{,}000 + 7{,}000(50) = 320{,}000$. The estimation may not be meaningful since 5000 is outside of the range of x values used to calculate the regression equation.

13.3

(a) See scatterplot below. The scatterplot shows a positive linear relationship.

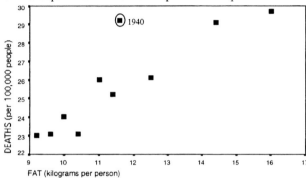

(b) $\hat{y} = 13.487 + 1.0649\, x$.

(c) $\hat{y} = 13.487 + 1.0649(13) = 15.331$ deaths per 100,000 people.

13.5

(a) A positive approximately linear relationship.

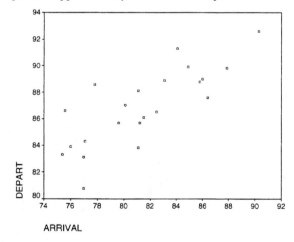

(b) $\hat{y} = 42.989 + 0.539\, x$.

13.7
(a)

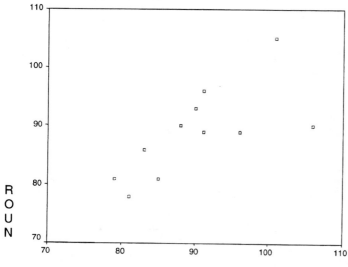

ROUND1

The observations 5 of (106, 90) could be influential since it is not consistent with the linear pattern and has a large x–value.

(b) $\hat{y} = 32.007 + 0.634\, x$.

13.9
Above, Below, Below, Above, Below, On.

13.11
(a) Negative. The slope of the regression equation $b = -10.877$.

(b) The regression equation is $\hat{y} = 137.083 - 10.877\, x$.

The predicted number of grievances is $\hat{y} = 137.083 - 10.877(8) = 50.067$.

(c) Yes. The p-value $= 0.001 < 0.05$. The data are statistically significant at the 0.05 level.

13.13
(c) $r = -0.77$.

13.15
(c) As one variable increases, the other variable tends to decrease.

13.17

Answers will vary.

(a)

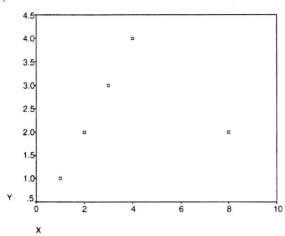

(b)

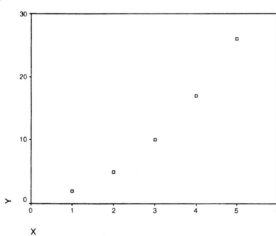

(c)

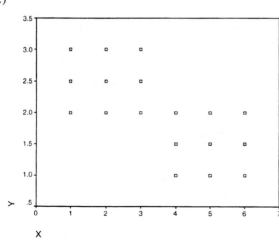

13.19
(a) 1
(b) 0

13.21
(a) $r = 0.908$.
(b) $\hat{y} = 6.863 + 0.862(30) = 32.72$.
(c) Unchanged.

13.23
$r = -1$.

13.25
(a) Silence interval before response.
(b) Number of words used in the response.
(c) $\hat{y} = 70.837 - 0.282\,x$
(d) $r = -0.194$.
(e) Stay the same.
(f) Since the p-value=0.646 > 0.10, we cannot reject the null hypotheses. There is no significant (non-zero) linear relationship between the two variables.

13.27
(a)

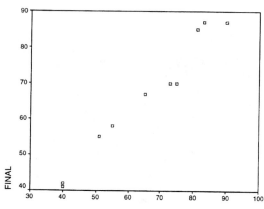

(b) $\hat{y} = 4.8595 + 0.9394\,x$.
(c) $\hat{y} = 4.8595 + 0.9394(70) = 70.6175$.
(d) residual $= e = 70 - 73.436 = -3.436$.

13.29

(a) Scatterplot

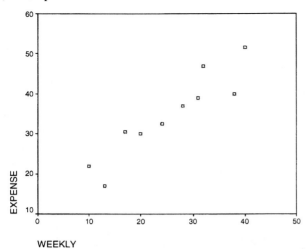

(b) $\hat{y} = 10.528 + 0.953\,x$.

(d) $\hat{y} = 10.528 + 0.953(21) = \$3,055$ in annual maintenance expenses.

(e) $r = 0.9252$.

(f) Residual Plot

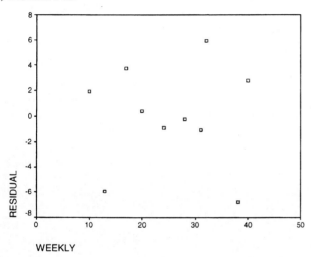

(g) Yes, the residual plot shows a random scatter in a horizontal band around 0.

13.31

(b) Expect the correlation to be positive, but not close to 1.

(c) $r = 0.565$.

(d) No, a truism of statistics is "correlation, not causation".

(e) Unchanged.

(f) $r = 1$.

(g) 0.682

(h) (i) Unchanged.

    (ii) 1.73

13.33

This may be a consequence of the regression effect. The volunteers will probably have higher blood pressure on average than normal people.

13.35

(a) $r = 0.967$. R Square = 0.936 means that 93.6% of the total variation in vein blood flow rate can be explained by the linear regression between sphere blood flow rate and vein blood flow rate.

(b) $\hat{y} = 1.031 + 0.902\,x$.

(c) $[0.902 \pm 2.306(0.083)] = [0.902 \pm 0.1924] = [\,0.711\,,\,1.093\,]$.

(d) $H_0: \beta = 1$ versus $H_1: \beta \neq 1$. Since 1 is inside the confidence interval we cannot reject H_0 at the 0.05 level.

(e) $\hat{y} = 1.031 + 0.902\,(12) = 11.855$.

(f) $[11.855 \pm 2.306\,(1.7566)\sqrt{1 + 1/10 + \dfrac{(12 - 13.07)^2}{443.2}}\,] => [7.602,\,16.108\}$.

(g) Narrower.

13.37

(a) IQ = 183.994 + 0.04222 (Weight) − 3.909 (Height) + 0.0002148 (MRIcount).

(b) R Square = 0.533. About 53.3% of the variability in IQ can be explained by the explanatory variables Weight, Height, and MRIcount in the multiple linear regression model.

(c) The estimate of the population standard deviation is 18.46.

(d) Each T-test statistic is for testing whether the particular regression coefficient is equal to zero. The p-values for Height and MRIcount are < 0.05, so they are statistical significant at the 0.05 level. Weight however is not statistically significant at the 0.05 level.

13.39

We need both variables to be quantitative. Gender is qualitative.

13.41

(a) There is a strong positive linear relationship since $r = 0.987$ is close to 1.

(b) $\hat{y} = 4.162 + 15.509(2) = 35.18$.

(c) Residual = e = 29 − 35.18 = − 6.18.

(d) Unchanged.

(e) Outlier.

13.43

(a)

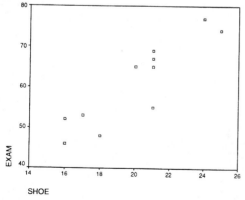

(b) $\hat{y} = -1.781 + 3.145\,x$.

(c) For a one shoe size increase, we would expect an exam score increase of 3.145 points on average.

(d) $\hat{y} = -1.781 + 3.145(45) = 70.554$ points.

(e) Age (or grade level).

13.45

(a) Predicted earnings $\hat{y} = 375337.492 - 11940.924(10) = \$255,982.25$.

(b) $r = -0.765$.

(c) (i) Unchanged.
    (ii) Unchanged.

13.47

(a) $r = -0.857$.

(b) $\hat{y} = 20.56 - 0.399\,x$.

(c) $\hat{y} = 20.56 - 0.399(42.2) = 3.72$.

(d) $120,000 = (1.2)(100,000)$ so we expect to have $(1.2)(3.72) = 4.46$ that is 4 or 5 cases of melanoma.

(e) No, 61.1 is outside the range of the x – values used to calculate the regression equation.

(f) (i) $r = 0.857$.

    (ii) $\hat{y} = 16.57 + 0.399\,x$.

(g) Residual Plot

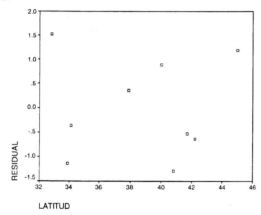

(h) Yes, the residuals form a random scatter around the zero-line.

13.49

(a) False.

(b) False.

13.51

(a)

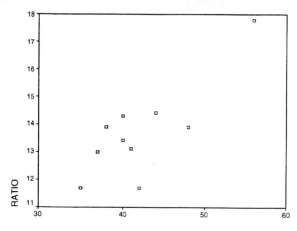

SALARY

(b) $\hat{y} = 4.08 + 0.2289\,x.$

(c) Residual $= e = 11.7 - 13.697 = -1\ 997.$

(d) $(42.1, 13.72);\ \hat{y} = 4.08 + 0.2289(42.1) = 13.72.$

(e) Yes. $(56, 17.8)$ (iii) Removing this observation it would markedly change the regression line.

13.53

(a) $\hat{y} = 321.241328 - 5.202654x.$

(b) The estimated slope of b= -5.202654 tells us that for one kg/day increase in total milk production for each cow, we would expect, on average, a decrease of 5.202654 uric-concentration in the milk.

(c) Yes. The p-value $= 0$ is given by Sig T.

13.55

(a) $\hat{y} = 70.717499 - 0.115611x.$

(b) 60 years old.

(c) $r = -0.80381.$

(d) Always.

(e) The p-value for assessing if there is a significant non-zero linear relationship is given in the output as the .0000 in the row with variable PERTV. Since this is so small we would conclude there appears to be a significant linear relationship between life expectancy and number of people per TV.

13.57

(a) Negative.

(b) $\hat{y} = 23.354 - 8.168\,x.$

(c) Residual $= e = 15 - 18.126 = -3.126.$

(d) The 95% confidence interval for β is: $1.2 \pm (0.20)(5.94) \Rightarrow (0.012, 2.388)$

(e) Using formula from page 810, we have:

$\hat{y} = 23.354 - 8.16\,(\,0.750) = 17.234.$

(f) $\sqrt{7.043\left(\dfrac{1}{12} + \dfrac{(0.75-1.13)^2}{1.66}\right)} = 1.095$

(g) 61.1%.

14.1
(a) 7.779
(b) 9.488
(c) 13.277
(d) 1.064
(e) 10.645
(f) 18.549

14.3
(a) With the TI: p-value = 0.3711; Using Table VI: $0.10 < p$-value < 0.90.
(b) With the TI: p-value = 0.0845; Using Table VI: $0.05 < p$-value < 0.10.
(c) With the TI: p-value = 0.0005; Using Table VI: p-value < 0.005.

14.5
(a) $H_0 : p_A = p_B = p_C = p_D = p_E = 0.20$
(b) The five expected counts are all $(200)(0.20) = 40$.
(c) $X_{OBS}^2 = \dfrac{(38-40)^2}{40} + \dfrac{(44-40)^2}{40} + \dfrac{(56-40)^2}{40} + \dfrac{(37-40)^2}{40} + \dfrac{(25-40)^2}{40} = 12.75.$
(d) With 4 degrees of freedom, the p-value is $P(X^2 \geq 12.75) = 0.0126$ (using the TI),

   or $0.01 < p$-value < 0.05 (using Table VI).
(e) Since the p-value is less than 0.05, we reject H_0 and conclude that the five pain relief medicines do not appear to be equally effective.

14.7
(a) $H_0: p_1 = 0.25 \quad p_2 = 0.50 \quad p_3 = 0.25$
(b) $E_1 = E_3 = (0.25)(144) = 26$ and $E_2 = (0.50)(144) = 72$

   $X^2 = 2.25 + 0.125 + 1 = 3.375$
   df = 2; $0.15 < p$–value < 0.20 (or p–value = 0.185 using the TI)
   So we accept H_0 and conclude that the professor's theory appears to be supported by the data.

14.9

The observed test statistic is $X_{OBS}^2 = \dfrac{(16-19.2)^2}{19.2} + \dfrac{(23-21.6)^2}{21.6} + \dfrac{(9-7.2)^2}{7.2} = 1.074$. With 2 degrees of freedom, the p-value is $P(X^2 \geq 1.074) = 0.5845$ (with the TI) or $0.10 < p$-value < 0.90 (using Table VI). Since the p-value is greater than 0.01, so we accept H_0 and conclude the distribution for domestic cars appears to be consistent with the distribution for non-domestic cars. Note: Since we observed a test statistic of 1.074 which is even less than what we'd expect to see if H_0 is true (df=2), it is certainly not large enough to reject H_0.

14.11
(a) About 26% of 1993 or 518.
(b) About 10% of 1993 or 200.
(c) Yes, at most 200.
(d) No, the categories are not disjoint (mutually exclusive), the percentages total to over 100%.
(e) The minimum sample size required for a 95% confidence interval with an error margin of 2% is 2401, and for 90% is 1692. The sample size of 1993 is between these two levels.
(f) No, we do not know anything about how the sample was obtained and where the adults are from.
(g) Is it a random sample? Where was the survey conducted? Were the adults all members of some health club? What kind of survey was it?

14.13

(a) The expected count for standard drug patients showing no change: $\dfrac{(100)(35)}{210} = 16.67$.

(b) The observed test statistic is 6.51. With 2 degrees of freedom, the p-value is $P\left(X^2 \geq 6.51\right) = 0.0386$ (with the TI) or $0.025 < p\text{-value} < 0.05$ (using Table VI). Thus our decision is to accept H_0 and conclude that the distribution for patient condition appears to be the same for the two treatments.

14.15

(a) H_0: The distribution of salary for union members is the same as the distribution of salary for non-union members.

H_1: The distribution of salary for union members is not the same as the distribution of salary for non-union members.

(b) $Expected = \dfrac{20(30)}{70} = 8.57$

(c) $X^2(2)$ distribution.

(d) Test statistic: 14.998; p–value = 0.001; Reject H_0 and conclude the distribution of salary does not appear to be the same for union members and non-union members.

14.17

(a) H_0: The three populations are homogeneous with respect to the distribution of alcohol usage, versus H_1: The three populations are not homogeneous with respect to the distribution of alcohol usage.

(b) Reject H_0 if the p-value is less than 0.05.

(c) You need to know the sample sizes.

(d) $X_{OBS}^2 = \dfrac{(34-50)^2}{50} + \dfrac{(68-50)^2}{50} + \dfrac{(66-50)^2}{50} + \dfrac{(52-50)^2}{50} + \dfrac{(32-50)^3}{50} = 23.36$.   The p-value is $P\left(X^2 \geq 23.36\right) = 0.0000085$ (with the TI) or $p\text{-value} < 0.005$ (using Table VI). Thus we reject H_0 and conclude that the distribution of alcohol usage does not appear to be the same for the 3 populations.

14.19

(a) H_0: Big toe status and Roll tongue status are independent versus H_1 Big toe status and Roll tongue status are not independent.

(b) The expected number people is $\dfrac{50(60)}{100} = 30$.

(c) The observed test statistic is 6.00, with a p-value of 0.01431. So our decision is to reject H_0. There does appear to be an association between Big Toe Status and Roll Tongue Status.

14.21

(a) H_0: Union membership and attitude are independent.

(b) The expected count for union members who are opposed is: $\dfrac{(176)(86)}{400} = 37.84$.

(c) The observed test statistic is 18.54498, with a p-value of 0.00009. Thus our decision is to reject H_0 and conclude that it appears that membership and attitude are associated.

(d) (i) H_0: Homogeneity, that is, the distribution of attitude is the same for the two populations.

(ii) We would reject H_0 since the p-value is less than 0.01.

14.23

(a) Proportion of men admitted: $90/200 = 0.45$ or 45%. Proportion of women admitted: $60/200 = 0.30$ or 30%. There appears to be an association between gender and admission status. The chi-square test of independence would give a test statistic of 9.6 and a p-value of 0.0019 (or using Table VI: p-value < 0.005). So there is strong evidence of an association between the two variables.

(b) Both departments admitted equally women and men applicants (50% in the engineering department, but only 25% in the English department). There does not appear to be any association between gender and admission status for either program. In fact the test statistic value for both programs would be 0 and the p-value $= 1$.

(c) Simpson's paradox or aggregation bias.

14.25

(a) The promotion rate for females of 260/1000 (or 26%) is higher than the promotion rate for males of 120/1000 (or 12%). A chi-square test homogeneity would yield an extremely large test statistic value of 63.68 and a p-value that is nearly 0. There is very strong support to say the incidence of promotion is not the same for male and female employees.

(b) Yes, as shown in the following possible example where females have a lower promotion rate at both job levels, but a higher proportion of females are in the job level the higher promotion rates overall.

	#	**Job Level 1** % promoted	# promoted	#	**Job Level 2** % promoted	# promoted
Males	100	30%	30	900	10%	90
Females	900	28%	252	100	8%	8

14.27

(a) H_0: $p_1 = 0.25$, $p_2 = 0.50$, $p_3 = 0.25$, H_1: At least one of the equalities is not true.
where p_1 = proportion of the resulting offspring "AA" , p_2 = proportion of the resulting offspring "AB", p_3 = proportion of the resulting offspring "BB".

(b) We would expect about $0.25(188) = 47$.

(c) With 2 degrees of freedom, the p-value is $P\left(X^2 \geq 3.2\right) = 0.202$ (with the TI) or $0.10 < p\text{-value} < 0.90$ (using Table VI). So, we fail to reject H_0 for any $\alpha > 0.202$. These data do appear to support Mendel's theory.

14.29

(a) $H_0 : p_1 = \dfrac{1}{3}, p_2 = \dfrac{1}{3}, p_3 = \dfrac{1}{3}$ or $p_1 = p_2 = p_3 = \dfrac{1}{3}$

(b) $E_1 = E_2 = E_3 = (120)\left(\dfrac{1}{3}\right) = 40$

$$X^2_{OBS} = \frac{(31-40)^2}{40} + \frac{(47-40)^2}{40} + \frac{(42-40)^2}{40} = \frac{81+49+4}{40} = 3.35$$

Using TI: the p-value $= 0.1873$. Using Table VI: $0.15 < p\text{-value} < 0.20$.
So we accept H_0.

14.31

(a) $H_0 : p_W = 0.61, \ p_A = 0.24, \ p_H = 0.06, \ p_B = 0.05, \ p_O = 0.04$

(b) We need to know the total number of students selected, that is, the sample size.

(c) The sample size is approximately $n = \dfrac{1127}{0.055} = 20{,}490$.

(d) The table of observed and expected counts is:

	White	Asian American	Hispanic	Black	Other
Observed	6557	7991	2828	1127	1987
Expected	12,498.9	4917.6	1229.4	1024.5	819.6

(e) The observed test statistic is $X_{OBS}^2 = 8497$, with 4 degrees of freedom and a p-value that is practically 0. Thus we reject H_0 and conclude that the model from 1984 does not appear to hold.

14.33

(a) H_0: The potency acceptance rates are the same for the 2 formulations, or the 2 formulations have homogeneous potency acceptance rates.

(b) The observed test statistic is 8, with a p-value of 0.00468 (with the TI) or p-value < 0.005 (using Table VI). Thus we reject H_0 and conclude that there appears to be a significant difference in the rates, with formulation 2 having the higher acceptance rate.

14.35

(a) The completed table is given by:

		District Type		
		Rural	*Suburban*	*Urban*
Goal	*Good Grades*	38.3%	57.6%	68.6%
	Popularity	33.6%	27.8%	17.1%
	Athletic Ability	28.2%	14.6%	14.3%
		100%	100%	100%

(b) The conditional distribution of Goal given District Type.

(c) Yes, the percentage with a goal of good grades increases from about 39% to nearly 69% as you go from rural to urban.

(d) $X_{OBS}^2 = 18.56$ is the observed test statistic with 4 degrees of freedom, the p-value is $P(X^2 \geq 18.56) = 0.00096$ (with the TI) or p-value < 0.005 (using Table VI). So our decision is to reject the null hypothesis, which means there appears to be an association between district type and goal.

14.37

(a) The conditional distribution of smoking status given treatment group is:

		Treatment Group		
		Group 1	*Group 2*	*Group 3*
Smoking Habit	*Nonsmoker*	19.8%	25.0%	19.2%
	Ex-Smoker	27.1%	24.0%	24.9%
	Smoker	53.1%	51.0%	55.9%
		100%	100%	100%

(b) Profile plot:

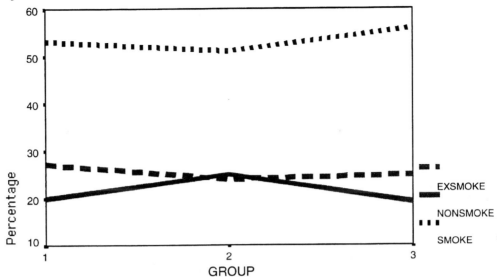

(c) The overall percentages are 21.4% nonsmokers, 25.3% ex-smokers, and 53.5% smokers.

(d) Enhanced profile plot:

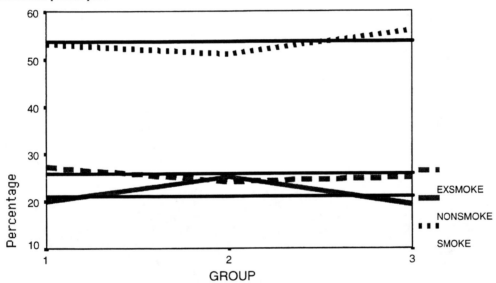

(e) Yes, the smoking habit rates are quite stable across the three groups.

Interactive Statistics 2nd Edition: Chapter 15 Odd Solutions

15.1
(a) $R_{OBS} = 7$; p-value $= 0.1172 + 0.0439 + 0.0098 + 0.0010 = 0.1719 > 0.05$, accept H_0.
(b) $R_{OBS} = 7$; p-value $= 0.9453 > 0.05$, accept H_0.
(c) $R_{OBS} = 7$; p-value $= 2(0.1719) = 0.3438 > 0.05$, accept H_0.
(d) The sample is selected randomly from a continuous population.

15.3
$$H_0 : \pi_{0.5} = 258 \text{ versus } H_1 : \pi_{0.5} > 258$$
$R_{OBS} = 10$; p-value $= 0.0916 + 0.0417 + 0.0139 + 0.0032 + 0.0005 + 0.00003 = 0.15093 > 0.05$ so accept H_0. The results are not statistically significant at the 5% level. There is not sufficient evidence to say the median round-trip ticket cost has increased.

15.5
(a) $H_0 : \pi_{0.5}(populationK) = \pi_{0.5}(populationG)$ versus
$H_1 : \pi_{0.5}(populationK) > \pi_{0.5}(populationG)$
(b) $W = 5 + 9 + 11 + 12 + 14 + 15 = 66$
(c) 48
(d) 8.49
(e) 2.12 standard deviations above
(f) 0.018
(g) The claim made by broker G, that its homes spend less time on the market, is supported by the data.

15.7
(a)

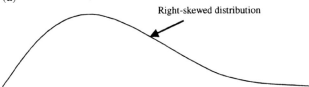

Right-skewed distribution

(b) $H_0 : \pi_{0.5}(population1) = \pi_{0.5}(population2)$ versus
$H_1 : \pi_{0.5}(population1) > \pi_{0.5}(population2)$ where 1 = Twin and 2 = Single.
(c) $W = 5 + 6 + 8 + 10 + 11 = 40$; p-value $= 0.041$
(d) Reject H_0.
(e) (i) Equal population standard deviations $\sigma_1 = \sigma_2$
(ii) $t(9)$ distribution

15.9
$$H_0 : \pi_{0.5}(populationA) = \pi_{0.5}(populationB) \text{ versus}$$
$$H_1 : \pi_{0.5}(populationA) \neq \pi_{0.5}(populationB).$$
$W = 66$; p-value $= 2(0.059) = 0.118$
Accept H_0 and conclude that there is not a significant difference between the median scores for the two schools at the 10% level.

15.11
$$H_0 : \pi_{0.50}(D) = 0 \text{ versus } H_1 : \pi_{0.50}(D) > 0 \text{ where } D = \text{December } - \text{January}$$
$W^+ = 46$ and the p-value $= 0.032$ so we reject H_0. There is sufficient evidence to say the median number of baggage-related complaints has decreased overall.

15.13

(a) $H_0 : \pi_{0.50}(D) = 0$ versus $H_1 : \pi_{0.50}(D) \neq 0$ where D = without − with

(b) $W^+ = 11$ and the p-value = 0.013 so we reject H_0.

15.15

$H_0 : \pi_{0.5} = 18$ versus $H_1 : \pi_{0.5} \neq 18$

$R_{OBS} = 2$; p-value = $2P(R \leq 2 \mid H_0$ is true$) = 2(0.1094 + 0.0313 + 0.0039) = 0.2892 > 0.05$ so we accept H_0. The results are not statistically significant at the 5% level. The data do not refute the claim that the median percentage of chromium present is 18%.

15.17

(a) $t = 3/1.5 = 2$

(b) $df = 18$, p-value is 0.0608 or from Table IV: $0.05 < p$-value < 0.10.

(c) Wilcoxon rank sum test.

(d) $\mu_W = \dfrac{n_1(n_1 + n_2 + 1)}{2} = \dfrac{10(21)}{2} = 105$

15.19

(a) Signed rank test

(b) $W^+ = 3 + 4 + 5 + 6 = 18$ and the p-value = 0.078

(c) Reject H_0

(d) The Paris Weekend seems to be more popular than the Beach Package.